Researching Institutions in Natural Resource Governance

This book presents an overview of different data collection methods and approaches that have been used to identify and analyse institutions associated with natural resource governance.

Institutions as rules of the game are fundamental to the management, governance and use of renewable natural resources, such as fisheries, forests and water. Yet researching institutions, particularly informal institutions, can be challenging as they can be difficult to identify, investigate and understand. This volume tackles this challenge head on. The innovative collection brings together contributions from different disciplines and traditions reporting and reflecting on a range of qualitative and quantitative research methods and experiences. This book begins by reviewing definitions and typologies of institutions and introduces different approaches to researching institutions in natural resource governance, namely the Bloomington School, associated with Elinor Ostrom, and Critical Institutionalism. Following this, each chapter provides an overview and example of application of research approaches and methods, including mixed methods, institutional grammar, fuzzy cognitive mapping, network analysis, assemblage and ethnography. The concluding chapter identifies key insights into institutions and lessons on researching institutions. From these, the chapter provides a set of observations on researching institutions in natural resource governance, with the aim of providing guidance for new and experienced researchers.

The volume provides an essential introduction for students and researchers to the range of methods and approaches, and their practical applications, for researching institutions associated with natural resource governance.

Fiona Nunan is a Professor of Environment and Development in the International Development Department at the University of Birmingham, UK. She is the author of *Understanding Poverty and the Environment* (Routledge, 2015), editor of *Governing Renewable Natural Resources* (Routledge, 2020) and lead editor of *The Routledge Handbook on Livelihoods in the Global South* (Routledge, 2022).

Earthscan Studies in Natural Resource Management

Resource Communities
Past Legacies and Future Pathways
Kristof Van Assche, Monica Gruezmacher, Lochner Marais, and Xaquin Perez-Sindin

English Urban Commons
The Past, Present and Future of Green Spaces
Christopher Rodgers, Rachel Hammersley, Alessandro Zambelli, Emma Cheatle, John Wedgwood Clarke, Sarah Collins, Olivia Dee and Siobhan O'Neill

Governing Natural Resources for Sustainable Peace in Africa
Environmental Justice and Conflict Resolution
Edited by Obasesam Okoi and Victoria R Nalule

Capacity-Building and the Water-Energy-Food Nexus
Rethinking Integration in the Asia-Pacific
Maureen Papas

Governing the Palm Oil Industry
Perspectives from Southeast Asia and Latin America
Edited by Patrick O'Reilly and Helena Varkkey

Researching Institutions in Natural Resource Governance
Methods and Frameworks
Edited by Fiona Nunan

Sustainable Management of Cordyceps
Supply Chains and Resource Management Policies
Edited by Jiping Sheng and Ksenia Gerasimova

For more information about this series, please visit: https://www.routledge.com/Earthscan-Studies-in-Natural-Resource-Management/book-series/ECNRM

Researching Institutions in Natural Resource Governance

Methods and Frameworks

Edited by
Fiona Nunan

Routledge
Taylor & Francis Group
LONDON AND NEW YORK

First published 2025
by Routledge
4 Park Square, Milton Park, Abingdon, Oxon OX14 4RN

and by Routledge
605 Third Avenue, New York, NY 10158

Routledge is an imprint of the Taylor & Francis Group, an informa business

British Library Cataloguing-in-Publication Data
A catalogue record for this book is available from the British Library

ISBN: 978-1-032-62466-2 (hbk)
ISBN: 978-1-032-62468-6 (pbk)
ISBN: 978-1-032-62467-9 (ebk)

DOI: 10.4324/9781032624679

Typeset in Galliard
by codeMantra

Contents

Figures

Tables

Contributors

Tavengwa Chitata is a Lecturer and Researcher at IHE Delft Institute for Water Studies, Netherlands. His work is at the intersection of irrigation engineering and social science. He uses interdisciplinary approaches to explore how people learn from and make sense of (ground)water through their everyday interactions with irrigation infrastructure. In his work, he pays explicit attention to the materiality of water infrastructure and how these shape flows of water and social relations of power, and vice versa.

Frances Cleaver is a Professor of Political Ecology at the Lancaster Environment Centre, Lancaster University. Her research is concerned with how we can understand natural resource governance in order to inform progressive social change. To this end she has written a book *Development through bricolage: Rethinking institutions for natural resource management* (Routledge 2012) and numerous articles and chapters developing ideas about institutional bricolage within a school of critical institutional studies. Working from a political ecology perspective, Frances is particularly interested in how institutions shape the governance of water, land and forests, and impact on people's livelihoods.

Sumana Datta is a Delhi-based environmental social scientist. Her work explores the intersection between environmental governance and rural development, with a particular focus on inclusive forest-carbon/climate governance and natural resource management. She has contributed chapters in books, written newspaper articles and published in prominent journals such as the *Economic and Political Weekly* and *Conservation and Society*. Previously, she served as an Assistant Professor at Ambedkar University Delhi and worked with Indian think tanks. She holds a PhD from the University of Manchester.

Graham Epstein is a Research Assistant Professor in the School of Environment, Resources and Sustainability at the University of Waterloo and Co-Director of Blue Range Solutions, a non-profit consultancy. His research addresses a range of questions at the intersection of people, the environment and governance to advance understanding of how policies and programs can be designed to support collective action, compliance and sustainable natural resource management and how complex social-ecological landscapes shape the design of institutions and influence their performance.

Tanya Hayes is a Professor at Seattle University where she chairs the Department of Public Affairs and Nonprofit Leadership and directs the Environmental Studies programme. She earned a joint PhD in Political Science and Public and Environmental Affairs from Indiana University under the supervision of Elinor Ostrom. Her research focuses on how conservation policies crafted in international or national policy arenas interact with communal governance structures, and in turn shape individual behaviours and social and ecological outcomes.

Jeltsje Kemerink-Seyoum is an Associate Professor of Water Governance and Justice at the IHE Delft Institute for Water Education. She is an interdisciplinary scholar who engages with critical theory to understand how injustices become manifest and how they can be challenged in and around water and agriculture. In particular, she studies the everyday practices of engineers, operators and irrigators who engage with water infrastructure to shed light on how the form and materiality of this infrastructure opens up space to renegotiate social relations of power. In her work, Jeltsje explicitly questions structures and hierarchies, including those based on gender, race and class.

Michael Kriegl is a marine social-ecologist and network scientist fascinated by the connections between people and the ocean. He wants to understand how social and ecological systems interact and what this means for the management of marine resources. At the Leibniz Centre for Tropical Marine Research (ZMT), he uses social-ecological network models to investigate the dynamics of small-scale fisheries in Latina America. Beyond his research, Michael is passionate about science communication, promoting ocean literacy and sparking curiosity in young minds.

James Douglas Langston is a landscape transition scientist at CSIRO. He completed a PhD at James Cook University on the Political Ecology of Spatial Development Initiatives, working with numerous partners in Indonesia. He managed a Masters in International Forestry programme out of UBC. Now he supports peatland restoration, establishing pathways to close gaps between different values demanded of peatlands. He enjoys the boundary work of landscape transition science and is motivated to reduce injustices that shape peoples' ability to prosper with nature.

Jelto Makris is an early-career researcher who studied International Relations and Political Science at Leiden University, Sciences Po Paris, and Free University Berlin. His research interests include ocean governance, resource conflicts in ocean spaces, and the intersection between environmental concerns, social justice and economic development, particularly in Latin America. He has worked on natural resource governance issues at the Food and Agriculture Organization (FAO), the Leibniz Centre for Tropical Marine Research (ZMT) and the Transnational Institute Amsterdam (TNI).

Candace K. May is a Rural and Natural Resources Sociologist with the United States Environmental Protection Agency. She examines rural people and places, natural resources governance, and the role of power and equity in access, use and allocation of resources. Candace employs evaluation and assessment techniques to understand how we can move towards environmental justice in contexts of socio-ecological complexity.

Felipe Murtinho is an Associate Professor at Seattle University. He is the Director of the International Studies Program and has an associate appointment with the Department of Public Affairs and Nonprofit Leadership and the Environmental Studies programme. He completed his Ph.D. in Geography at University California Santa Barbara. His research aims to understand how rural communities in Latin America impact their environment and how they collectively respond to changes in the natural resource systems that they depend on for their livelihoods.

Ben Nagel is a researcher in the working group Institutional and Behavioural Economics at the Leibniz Centre for Tropical Marine Research (ZMT) in Bremen, Germany, where he recently completed his PhD in Political Science (2024) in conjunction with Constructor University. His research focuses on social-ecological systems analysis of sustainability transitions in food production systems including agriculture and aquaculture. His recent work explores how advances in network analysis and participatory modelling can bridge science and society and inform sustainability transformations.

Fiona Nunan is a Professor of Environment and Development in the International Development Department, University of Birmingham, researching collaborative governance of renewable natural resources, particularly inland fisheries in East and Southern Africa. She is the author of *Understanding Poverty and the Environment: Analytical frameworks and approaches*, editor of *Making Climate Compatible Development Happen* and *Governing Renewable Natural Resources: Theories and frameworks* and lead editor of *The Routledge Handbook on Livelihoods in the Global South.*

José Carlos Orihuela studies how institutions evolve and matter for economic development and environmental governance in Latin America. His research has appeared in the *Journal of Institutional Economics*, *Studies in Comparative International Development*, *Forest Policy and Economics*, *New Political Economy*, among other outlets. He co-edited *Resource Booms and Institutional Pathways: The Case of the Extractive Industry in Peru* (Palgrave 2017) and co-authored *The Developmental Challenges of Mining and Oil: Lessons from Africa and Latin America* (Palgrave 2012).

Rebecca Anne Riggs is a sustainability scientist at CSIRO. Before joining CSIRO, Rebecca was a postdoctoral research fellow at the University of British Columbia in Canada after completing her PhD at James Cook University in Cairns, Australia. Rebecca works in transdisciplinary research partnerships focused on strengthening capabilities to explore, understand and act on challenges of just sustainability transitions. Her work seeks to incorporate local values and knowledge systems into natural resource governance, with collaborations in Indonesia, Cambodia, Cameroon, Canada and Australia.

Mas Davino Sayaza (Vino) graduated from the Masters in International Forestry programme at the University of British Columbia in 2023 following his Bachelor degree in Communication Management at Universitas Padjadjaran, Indonesia. Vino works as a research assistant looking into environmental issues and a communication assistant at IUFRO Division 6. Vino's recent research utilizes the Sustainable Livelihood Framework to improve the lives of rural communities in Indonesia and offers technical and facilitation support to various research programmes within the country.

Edella Schlager is a Professor of Public Policy in the School of Government and Public Policy at the University of Arizona, Tucson. Her research centres on comparative institutional analysis with a primary focus on watershed and river basin governance. She is a member of the Institutional Grammar Research Initiative (institutionalgrammar.org). In 2022, she received the Elinor Ostrom Lifetime Achievement Award from the Science, Technology, and Environmental Politics Section of the American Political Science Association.

Achim Schlüter leads the working group of Institutional and Behavioural Economics at the Leibniz Centre for Tropical Marine Research in Bremen, Germany, and holds a professorship of Institutional Economics and Ocean Governance at the University of Bremen. He is interested in institutional change in rapidly emerging blue economies. Oceans and coasts contain numerous diverse common-pool resources. Nevertheless, we see a lot of forms of private property arrangements emerge. Understanding them and discussing holistically sustainable governance solutions are his main interests.

Saba Siddiki is a Professor in the Maxwell School of Citizenship and Public Affairs at Syracuse University. She is the Co-Director of the Institutional Grammar Research Initiative (institutionalgrammar.org). Her research expertise is in policy design, collaborative policymaking, institutional analysis, sustainability, environmental governance and food system governance.

Palden Tsering is an Associate Professor in the School of Ethnology and Sociology, Qinghai Minzu University, China, and Associate Editor for the journal of *Pastoralism: Research, Policy and Practice.* He has been engaged in the development, conservation and governance of resources, as well as the politics of these dynamic interactions amongst diverse institutions amid changes and uncertainties in the pastoralist context.

Ryan R. Unks is an interdisciplinary scientist whose work is primarily concerned with the relationship between pastoral livelihoods, dryland ecology and land governance. He has worked at the National Socio-Environmental Synthesis Center (US) and the University of Lyon 2 (FR), and holds a PhD in Integrative Conservation from the University of Georgia (US). He is currently part of the ERC funded Conservation Data Justice project based at the Institut de Ciència i Tecnologia Ambientals (ICTA-UAB) at the Universitat Autònoma de Barcelona (ES).

1 Theory and methods in the study of institutions

Fiona Nunan

Introduction

Institutions are frequently cited in relation to natural resource governance (Nunan, 2020). They are attributed with the capacity to influence, perhaps even determine, what governance looks like, how it performs, and what the outcomes of governance are. Very often multiple institutions exist within a governance system and its broader context, at multiple administrative levels, from local to international. Different types of institutions may co-exist but may also interact and change. Understanding what is going on in terms of which institutions are influencing governance, why and how, can therefore be challenging.

This book aims to address the challenges experienced in identifying, understanding and investigating institutions in natural resource governance, i.e. in researching institutions. Institutions are understood in this book as 'rules of the game' (North, 1990), with these rules affecting who is involved in governance, and having implications for decisions made and social and ecological outcomes. There are many research designs, methods and approaches that have been used to research institutions in relation to natural resource governance, with decisions made on which methods to use informed by theory, researcher preferences and the research questions being asked. There is no one best method to use or theory to draw on. So, the aim of this book is not to recommend certain methods, frameworks or approaches. Rather, the aim of the book is to share a range of experiences of using different methods, frameworks and approaches, and reflections by researchers on their experience. Descriptions and reflections on the experience of using a range of methods, frameworks and approaches should be of interest and benefit to other researchers, providing theoretical, practical and logistical lessons for future research. Each chapter introduces a method, framework or approach, explains why and how the authors used that method/framework/approach and in what context, what decisions they made during a research process, what challenges were encountered and whether and how those were overcome.

This chapter explains the purpose of the book, provides a rationale for focusing on natural resource governance, reviews definitions of institutions and introduces key approaches to understanding institutions in natural

DOI: 10.4324/9781032624679-1

resource governance. The final section of this chapter introduces the chapters that follow, in relation to the methods, frameworks and approaches described and reflected on, and the cases used to illustrate application and gain insights from. The book finishes with a concluding chapter that identifies key themes and lessons from across the chapters, providing an overall guide to planning for and undertaking research into institutions in natural resource governance.

Why natural resource governance?

Institutions exist and influence every part of our lives and are probably challenging to research in every context. This book focuses on natural resource governance; more specifically, the governance of renewable natural resources. This is for at least two reasons. The first is that in recognition of the fundamental role of institutions in natural resource governance, many researchers undertake some form of institutional analysis to investigate a range of research problems (Nunan, 2020). This means that many methods, frameworks and approaches have been, and could be, used. This range of experience gives us many examples to learn from, but it also means that it can be challenging for researchers, particularly those new to institutional analysis of natural resource governance, to know which methods might be more appropriate to use in a given situation. The second reason concerns the number and diversity of institutions found within and beyond natural resource governance. Institutions may be crafted for the purpose of natural resource governance or management, but they may also exist independently of natural resource governance but influence it. That may be through state institutions, government policy or through customary systems. Institutions are also considered to emerge and evolve over time, and to interact, and change as a result of interaction.

Reasons for the number and diversity of institutions operating within and influencing natural resource governance include the characteristics of renewable natural resources themselves. The term 'renewable natural resources' refers to resources such as fisheries, forests and grazing land, which are considered able to replenish themselves, as opposed to non-renewable natural resources such as minerals and oil. Of course, there are limits to the renewability of natural resources, as reflected in the categorisation of these resources as 'common pool'. 'Common pool resource' is a term originating in economics that refers to resources with two defining characteristics, of high costs of exclusion and high levels of subtractability. 'High costs of exclusion' imply that it is difficult to stop people from using or extracting the resource, whereas as subtractability refers to the consumption of the resource by a user reducing availability for others (Ostrom et al., 2002). These characteristics suggest that in managing the number and diversity of users of natural resources, and resulting competition, rules, or institutions, are needed.

Definitions of natural resource governance often make explicit reference to institutions. Crona et al. (2011: 45), for example, define natural resource governance in reference to distinguishing governance from management, as a 'broader system of formal or informal institutions in which the management actions are embedded and which provide the essential direction, resources, and structure needed to meet the overarching resource governance goals'. Campese (2016: 7) similarly recognises diversity in institutions within natural resource governance, defining it as

> the norms, institutions, and processes that determine how power and responsibilities over natural resources are exercised, how decisions are taken and how citizens – including women, men, youth, indigenous peoples and local communities – secure access to, participate in, and are impacted by the management of natural resources.

These definitions make clear the centrality of institutions to enacting governance. They also are clear in differentiating governance from management. However, the boundaries between governance and management can sometimes be blurred (Arts, 2014), so institutional analysis of governance may encompass aspects of management.

Institutional analysis of natural resource governance may not specifically refer to natural resources but to the flows of benefits in the form of 'ecosystem services'. The term 'natural resources' tends to be associated with economics and physical sciences, reflecting the extraction of nature and its management for the benefit of people, and often for profit. The concept of ecosystem services takes a more dynamic view of the benefits people receive from nature. The concept of ecosystem services originates in ecology but is also strongly associated with environmental economics and the valuation of nature. The governance of ecosystem services has received increasing attention, sharing principles and characteristics with that of natural resource and environmental governance (Tucker et al., 2023). Ecosystem services governance has been defined as 'the institutionalisation of mechanisms for collective decision-making and collective action with respect to natural resource management' (Rival and Muradian, 2013: 4), a very similar definition to those of natural resource governance.

In searching for examples of how institutions have been researched in relation to natural resource governance, cases may be found then that refer to ecosystem services. Literature may also refer to governance in relation to social-ecological systems (SESs) (Aggarwal and Anderies, 2023), most notably in relation to the Social-Ecological Systems framework (SESF) that builds on Elinor Ostrom's Institutional Analysis and Development (IAD) framework, as further discussed below. This book includes cases that refer to institutional analysis in diverse settings, including fisheries, grazing land, water and forests, referring to natural resources, ecosystem services and SES. Institutions are relevant, diverse and multifaceted whichever framing is used.

Understanding institutions

Much research into institutions within literature on natural resource governance refer to institutions as 'rules of the game' drawing on Douglas North's (1990) classic definition, which, more fully, refers to institutions as 'the rules of the game in a society, or more formally, are the humanly devised constraints that shape human interaction', which 'reduce uncertainty by providing a structure to everyday life' (North, 1990: 3). How institutions are defined matters for how they are understood, interpreted and researched. While North's (1990) definition emphasises how institutions reduce uncertainty and increase predictability, other definitions observe that this does not imply that institutions do not change. Crawford and Ostrom (1995: 582) describe institutions as

> enduring regularities of human action in situations structured by rules, norms, and shared strategies, as well as by the physical world. The rules, norms, and shared strategies are constituted and reconstituted by human interaction in frequently occurring or repetitive situations.

Institutions according to this definition are regularly re-formed, recognising the dynamic nature of institutions as well as their stability and defining capacity to enable predictability in decisions, behaviour and interactions.

As well as alluding to a more dynamic view of institutions, Crawford and Ostrom's (1995) definition was instrumental in identifying components or dimensions of institutions. These are set out as rules, norms and shared strategies, forming the basis of the grammar of institutions and of 'institutional statements', subsequently leading to the development of the institutional grammar tool (Lien, 2020). The grammar of institutions enables statements to be broken down into their component parts, with an institutional statement being described as 'a shared linguistic constraint or opportunity that prescribes, permits, or advises actions or outcomes for actors' (Crawford and Ostrom, 1995: 583). Institutional grammar enables systematic analysis by coding phrases within statements and bringing the coded phrases together into larger datasets (Basurto et al., 2010). How institutional grammar can be applied is further explained in Chapter 3, where Schlager and Siddiki provide detailed guidance on how to analyse institutional statements and use institutional grammar in conjunction with other analytical approaches.

Using institutional grammar to unpick institutional statements allows analysts to categorise institutions as rules, norms or shared strategies, as well as classify statements according to institutional type, of which Ostrom (2005) identifies seven. For the purposes of this chapter, institutional types are not reviewed, but definitions of rules, norms and shared strategies contribute to understanding the forms and diversity of institutions. For Ostrom (2005: 18), rules are 'shared understandings by participants about *enforced* prescriptions concerning what actions (or outcomes) are *required*, *prohibited*, or *permitted*' (original italics). In turn, Ostrom (2005: 112) describes norms as 'shared concepts of what must, must not, or may be appropriate actions or outcomes in particular types of situations'. Norms suggest 'moral behavior, ethical standards or patterned [conventional] ways of doing things'

(Yeboah-Assiamah, Muller and Domfeh, 2017: 2). Drawing on Ostrom (2005), Schlüter and Theesfeld (2010: 452) explain 'shared strategies' as 'behaviour of a player in a game' and note that strategies differ from rules and norms by the absence of sanctions.

Other perspectives on and approaches to institutional analysis of natural resource governance give greater emphasis to the dynamic nature of institutions, suggesting that institutions evolve rather than are designed. Cleaver (2012: 8), for example, defines institutions as 'arrangements between people which are reproduced and regularised across time and space and which are subject to constant processes of evolution and change'. This view of institutions is reflected in the analytical approach of 'critical institutionalism', which suggests that institutions emerge, interact and reflect historical processes and context (Cleaver, 2012; Hall et al., 2014).

As well as differentiating between rules, norms and shared strategies as different forms of institutions, institutions are also often categorised in a binary manner as either formal or informal. Drawing on Lauth (2000), Yeboah-Assiamah, Muller and Domfeh (2021: 2) conclude that formal institutions 'involve the prescriptions, instruments and instructions that are largely codified, having the status of constitutional clauses and laws that are guaranteed and sanctioned at multiple levels largely by public agencies'. The term 'informal institution' can be harder to attribute to certain forms or sources of institutions. Helmke and Levitsky (2004: 727) define informal institutions as 'socially shared rules, usually unwritten, that are created, communicated, and enforced outside of officially sanctioned channels' and caution that identifying and measuring informal institutions can be challenging. Po et al. (2019: 1333) recognise this challenge in relation to researching natural resource management, noting that informal institutions are 'often implicit' and 'difficult to observe'. Informal institutions of relevance to natural resource management are seen to include social taboos, as they have the capacity to protect habitats and ecosystems and be respected by local communities (Colding and Folke, 2001).

As well as informal institutions being difficult to identify and research, literature has also cautioned that the boundaries between formal and informal institutions can be blurred (Po et al., 2019). An example often cited is when rules associated with customary systems are recognised by the formal state system. Cleaver (2002) responds to the challenges associated with distinguishing between formal and informal institutions by employing the terms bureaucratic and socially-embedded institutions instead. These she defines as:

> Bureaucratic institutions are those formalised arrangements based on explicit organisational structures, contracts and legal rights, often introduced by governments or development agencies. Socially embedded institutions are those based on culture, social organisation and daily practice.
>
> Cleaver (2002: 13)

Cleaver's differentiation between bureaucratic and socially-embedded institutions is reflected in her development and application of critical institutionalism, discussed further below.

Whichever terminology is used, it is apparent that there are many forms and sources of institutions, and that it is not always easy or possible to distinguish between formal and informal institutions. As discussed later in relation to institutional bricolage, institutions, whether formal or informal, interact and change, reflecting the observation that when new institutions are developed and implemented, they are not introduced into an institutional vacuum. Many other institutions do, and did, exist in any context. In addition to institutional bricolage, processes of institutional bargaining and layering have been identified, referring to the emergence of new arrangements as a result of negotiation and bargaining, and incremental additions to existing institutions (Marin and Bjørklund, 2015). Institutional layering is considered a slow and piecemeal process, in contrast to sudden shocks to a system portrayed through theories such as punctuated equilibrium (Jupille and Caporaso, 2022). However, layering can bring about substantial changes in institutions if the logic of the institution changes or the capacity of the essence of the institution to be reproduced is constrained (Mahoney and Thelen, 2010).

Institutional history and interplay are reflected in Po et al.'s (2019: 1337) observation that the development of formal institutions may be multifaceted, being 'partly inherited, partly designed, partly evolved, and partly adapted from other rules'. In addition, Po et al. (2019: 1337) advise that formal institutions 'heavily depend on informal institutions structures to propose, interpret, and adapt rules to specific contexts' and that formal and informal institutions may '(co)evolve following diverse and often non-linear trajectories'. An analytical framework to investigate this dependence and co-evolution is provided by Helmke and Levitsky (2004), who identify four types of informal institutions in terms of how they engage with formal institutions. These four types are complementary, accommodating, substitutive and competing, referring to whether the informal institution seeks the same outcome as the formal institution and whether the informal institution encourages the same or conflicting behaviour as the formal institution. Osei-Tutu, Pregernig and Pokorny (2015), in applying this typology to forest management in Ghana, found that it is not always possible in practice to discern a consistent intention of institutions in practice. This makes it challenging to determine whether institutions are converging or diverging, suggesting that some institutions may encourage different behaviours and outcomes in various contexts and situations.

Institutional analysis can involve the use of many methods, seeking answers to a diversity of questions. Institutional analysis has been explained as being 'oriented around the role of institutions in shaping the incentives, opportunities and constraints that actors face as they interact with the environment and one another' (Epstein, Villamayor-Tomas and Schoon, 2021: 309). But institutions must be identified, potentially classified or categorised. In addition, research might investigate the performance of institutions, how they interact and how and why they change (Imperial, 1999). Much institutional analysis of natural resource governance has been inspired by and fall within the two main schools introduced already, which are now further reviewed. The first is the 'Bloomington School', referred to by Cleaver (2012) as 'mainstream institutionalism' in contrast to the 'critical institutionalism' of her own work, which is the second school of institutional analysis reviewed. Both mainstream and critical institutionalism fall within the broad remit of 'new institutionalism', which contrasts with 'old institutionalism' through greater recognition of informal institutions and challenging assumptions of rational decision-making, given that there is rarely complete information available (Lowndes and Roberts, 2013).

Bloomington School: mainstream institutionalism

The 'Bloomington School' refers to the work associated with the Ostrom Workshop, founded by Elinor and Vincent Ostrom and based at Indiana University, Bloomington, USA. Institutional analysis of the governance of common pool resources is particularly associated with Elinor Ostrom, greatly informed by her 1990 publication, *Governing the Commons: The Evolution of Institutions for Collective Action*. The work of the Ostroms employed a political economy approach and developed approaches to institutional analysis that recognised bounded rationality, as opposed to assuming that actors are all-informed and can make fully rational decisions.

Several key developments are associated with Elinor Ostrom's work on institutional analysis, including the 'design principles' identified from analysis of long-lasting common property regimes that reflect the ingredients needed for sustainable governance, and the IAD framework. The design principles reflect the influence of attributes of actors, institutions and common pool resources that affect potential sustainability. Table 1.1 sets out the design principles as articulated in Ostrom (1990). While the set and wording of design principles has been critiqued, they have been found to be consistently supported by empirical evidence (Cox, Arnold and Villamayor Tomás, 2010). Modifications of the design principles have been put forward, but the original set provide ongoing foundation for much institutional analysis of natural resource governance.

The IAD framework provides a tool to investigate and explain the ways in which 'institutions operate and change over time' (McGinnis, 2011: 169). The framework identifies a number of 'exogenous variables', being biophysical/material conditions, attributes of community and rules, which influence the 'action arena', in which there is interaction between action situations and actors. Action situations are described as 'the "black box" where policy choices are made' (McGinnis, 2011: 172) and as 'spaces of interdependent decision-making' (Epstein, Villamayor-Tomas and Schoon, 2021: 310). Interactions between the action situation and actors, or participants, in the action arena inform, and are informed by, interactions leading to outcomes, with

Table 1.1 Ostrom's design principles for sustained and effective common pool resource governance

Design principle
There are clear boundaries agreed, identifying the common pool resource and those who can benefit
Rules on extraction and benefits should be related to local conditions and context, such as when resources can be extracted and how much can be taken by whom
Those affected by operational rules should be able to participate in changing those rules through collective-choice arrangements
Monitoring should be undertaken of how the common pool resources are being used, with monitoring being accountable to those who have rights to extract
Graduated sanctions should be in place to address non-compliance, enabling sanctions to reflect the type and number of previous offences
Conflict resolution should be facilitated through having mechanisms in place
Those with rights under the common property regime should be able to make their own institutions without being challenged by government authorities
For common property regimes that are part of larger systems, these should be organised in multiple layers through nested systems to ensure institutions are consistent

Source: Adapted from Ostrom (1990: 90).

'evaluative criteria' assessing interactions and outcomes. Outcomes feedback into the exogenous variables and to the action arena, leading to additional interactions and outcomes. The IAD framework is further explained in Chapter 2, with guidance and reflections on its application. Resources that provide more detailed information on the exogenous variables, components of the action situation and evaluative criteria include Epstein, Villamayor-Tomas and Schoon (2021), McGinnis (2011) and Ostrom (2005 and 2011).

Of particular relevance to researching institutions, Ostrom (2005) identified four levels of rule-making and rules: metaconstitutional, constitutional, collective-choice and operational. Institutions can, therefore, be found at each of these levels and are interconnected. The metaconstitutional level provides the overarching framework for the rules and decision-making lower down, setting out the elements for the constitutional level. Constitutional rules determine how collective-choice and operational rules can be created and changed, and who can be involved. Collective-choice rules in turn guide operational rules by setting out how operational rules can be changed and who can participate in decision-making. Operational rules set out the detailed operational matters of commons governance in relation to what resources can be extracted, by whom, how much and when, how monitoring will take place and how activities that go against the rules will be sanctioned. Institutional analysis of natural resource governance could involve investigation of the formation and functioning of rules at all levels, and into how they interact or interconnect.

The multiple levels and nesting of rules in common property regimes are reflected in the concept and practice of 'polycentric governance' or polycentricity, a further area of study and theorisation associated with the

Bloomington School. Polycentricity refers to 'a complex form of governance with multiple centers of decision making, each of which operates with some degree of autonomy' (Carlisle and Gruby, 2019: 928). These multiple centres of decision-making are nested at administrative levels, such as local, sub-national and national, and include specialised governance agencies that cut across administrative levels. For a governance system to be considered polycentric, there should be coordination between the centres. Institutional analysis of natural resource governance may, therefore, involve analysis of multiple centres, their linkages and coordination, and how these relate to the resource itself. The relationship to the resource, or SES, is often assessed in terms of 'institutional fit', referring to the compatibility of resource regimes and biogeographical systems (Young, 2002). Institutional analysis may be employed to investigate how institutional arrangements reflect the scale and dynamics of ecosystems.

Building on the IAD framework, Ostrom and others subsequently developed the SESF, which builds on recognition that interdependent social and ecological variables affect institutional governance arrangements and outcomes (Ostrom, 2007, 2009; Partelow, 2018). The 'action situation' is central to the SESF as it is to the IAD framework, with four components of resource systems, resource units, governance system and actors comprising of multi-tiered sets of variables that interact and influence, and are influenced by, what goes on in the action situation. The main purpose of this framework, though aims of application vary, is to enable 'diagnosis' of what is happening in an SES, including how governance arrangements are structured and performing, recognising that there is no one perfect solution for common pool resource governance. Modifications of the SESF and efforts to build on the SESF have led to further frameworks and approaches, including several that aim to facilitate analysis of how institutions develop and change over time (Epstein et al., 2020).

Institutional analysis associated with the Bloomington School is noted for drawing on a range of methods, rather than relying on one specific method, and often drawing on multiple methods, or being 'mixed-method', drawn from a range of quantitative and qualitative methods (Labrousse, 2016; Partelow, 2018; Poteete, Janssen and Ostrom, 2010). Epstein, Villamayor-Tomas and Schoon (2021: 311–312) noted a range of methods that have been used, including 'case studies and ethnographic research, common-pool resource experiments…behavioural experiments…interviews'. Poteete, Janssen and Ostrom (2010) explained how some of these methods have been used in research into collective action for the management of natural resources, including case studies, meta-analysis and laboratory and field-based experiments. Case studies can be used to investigate causal relationships and develop concepts and theories, whereas experiments are used to test hypotheses and meta-analysis enables larger sets of data to be developed from a number of small-n studies by integrating and systematically coding data (Poteete, Janssen and Ostrom, 2010).

Critical institutionalism and institutional bricolage

Critical institutionalism offers a different approach to and perspective on analysing institutions in relation to natural resource governance. Critical institutionalism analysis of natural resource governance has been particularly informed by Cleaver (2002, 2012), whose work builds on critique of 'mainstream institutionalism' of the Bloomington School. Cleaver (2002: 11) argued that the 'institutional crafting' associated with common property theory 'is based on concepts which are inadequately socially informed and which ill-reflect the complexity, diversity and *ad hoc* nature of institutional formation'. Critical institutionalism does not, however, reject mainstream institutionalism, with Cleaver and Whaley (2018: 2) noting that the approach has 'one foot in commons scholarship' and Whaley (2018) bringing critical institutionalism together with the IAD framework to produce a 'Critical Institutional Analysis and Development Framework'.

The approach responds to this critique of mainstream institutionalism, being informed by insights from political ecology, sociological institutionalism and legal pluralism (Cleaver, 2012). The defining characteristics of critical institutionalism are set out as recognising '(i) complexity of institutions entwined in everyday social life; (ii) their historical formation; and (iii) the interplay between the traditional and the modern, formal and informal arrangements' (Hall et al., 2014: 73). A critical institutionalist approach is particularly characterised by recognition of the role and influence of institutions that were not designed as part of natural resource governance but are significant anyway. These may include power relations, gender relations and kinship. The approach has similarities to sociological institutionalism, where institutions are considered to be part of the wider political and cultural context, implying that understanding of the wider context is essential to analyses of natural resource governance (Sandström, 2009).

The second defining characteristic of a critical institutionalist approach is the attention given to analysing why, how and with what implications institutions interact, referring to such interaction as 'institutional bricolage'. Institutional interplay is recognised within many other traditions of institutional analysis, alluding to interplay with concepts such as legal pluralism and inter-institutional gaps (Po et al., 2019; Rahman et al., 2017). Definitions of institutional bricolage provide insight into how institutional interplay is observed to occur, with de Koning and Cleaver (2012: 281) offering the following definition: 'a process through which actors consciously and unconsciously reshape or piece together different arrangements at hand'.

de Koning (2014) identified several processes or practices of institutional bricolage, as aggregation, alteration and articulation, through research into how communities responded to the introduction of community forestry in the Amazon Region of Bolivia and Ecuador. She explains aggregation as referring to 'the recombination of newly introduced institutions and locally embedded institutions' (de Koning, 2014: 360); alteration as 'the adaptation of institutions. This can happen to both newly introduced institutions and locally embedded institutions. Local actors often tweak or tinker with forest regulations to make them better fit their livelihoods or identity' (de Koning, 2014: 360); and, articulation as 'the claiming of traditional identities and culture and the rejection of newly introduced institutions' (de Koning, 2014: 360). Institutional analysis involving institutional bricolage may, therefore, involve analyses of how institutions interact and change as a result of how communities respond to governance reform.

Karambiri et al. (2020) dig deeper into institutional interaction, interplay and bricolage by seeking to explain why communities respond in different ways to governance reform, leading to different institutional, social and ecological outcomes. They investigate three explanatory factors of agency, networks and power, to explain how the response and outcomes of three villages to the introduction of community forestry in Burkina Faso differ. They found that analysis of agency, networks and power explained how and why communities responded to governance reform as they did and concluded that greater recognition is needed of how local actors drive institutional change. From similar findings in Nepal, Haapala et al. (2016) suggested that bricolage could be anticipated from the outset of governance reform, with those having agency and power facilitating bricolage and acting as 'bricoleurs'.

Methods associated with critical institutionalism tend to be more qualitative than quantitative. de Koning (2014: 361) referred to using 'in-depth interviews, participant observations, group interviews, group exercises and questionnaires'; Funder and Marani (2015) undertook interviews and convened focus group discussions, with data collected over years, bringing a longitudinal dimension to their research into how Environment Officers navigated bureaucratic and socially-embedded institutions to undertake their work; and Kairu et al. (2018) referred to shadowing officers of the Kenya Forest Service, with participant observation and interviews undertaken to understand how forest officers at different administrative levels interacted with community forest governance. These approaches fall within ethnography, which enables deep engagement with a research context and participants to generate understanding of the nuances of institutions and institutional bricolage.

Conclusions

From this introduction to literature on institutions and natural resource governance, several conclusions can be drawn. The first is that institutions are consistently understood as 'rules of the game', with definitions recognising the role that institutions play in reducing transaction costs and bringing predictability to interactions. However, definitions also recognise that institutions may be re-formed during interactions, they may change over time and they interact with each other. Interaction between institutions results from a second observation from the literature that there are very often multiple institutions within any context, meaning that when a new institution is implemented, it does not enter an institutional vacuum. Historical context and existing institutions matter and may affect how a new institution is received, performs, and is shaped over time and space. Institutions may be categorised in different ways, though it also apparent that boundaries between categories can be blurred.

Approaches to understanding institutions and informing research differ in how they view the source of institutions, i.e., whether institutions can be crafted or designed, or whether they emerge, evolve and are pieced together from other institutions. Research may also find that institutions have not only been crafted and introduced but also shaped through their implementation and interaction with other institutions.

What is clear is that there are very often many institutions associated with natural resource governance and that some institutions, particularly informal (or socially-embedded) institutions, can be hard to identify, understand and research. In addition, there are many frameworks, approaches and methods that can, and have been, used to investigate institutions and natural resource governance. Learning from researchers' decision-making and experience in using a range of frameworks, methods and approaches may, therefore, assist others in their decision-making, research design and in data analysis and interpretation.

Structure of the book

Chapters 2–5 of the book are largely inspired by and founded in Ostrom's work on institutions, beginning with an example of application of the IAD framework, used in the case study to link collective action, behavioural change and social-environmental outcomes in conservation policy. Research into a Payment for Ecosystem Services scheme in Ecuador is used by Hayes and Murtinho to explain how the IAD framework and associated methods were used, providing insight in particular on taking a phased approach to research, researching institutions over time and triangulating data. In Chapter 3, Siddiki and Schlager further introduce and reflect on institutional grammar, providing detailed guidance and illustrations of this analytical tool. Chapter 4 takes a different perspective, addressing challenges experienced in measuring the effectiveness of institutions. Epstein reviews five approaches to measuring institutions, focusing on the content, inputs, activities, outputs and outcomes associated with institutions, and identifies key methods used, and limitations found, in relation to these approaches. In Chapter 5, May introduces the complex adaptive governance systems framework, which combines the panarchy heuristic and Ostrom and colleagues' institutional levels with compatible theories of human practices, as a tool for research into complex institutional systems, recognising diversity of identities and interests.

The following two chapters, Chapters 6 and 7, take a mapping or network approach to identifying and investigating institutions. In Chapter 6, Makris et al. introduce fuzzy cognitive mapping as a participatory modelling method used to create causal maps with, or by, research participants, which can be used to compare their experiences of, and perspectives on, institutions. Riggs, Langston and Sayaza in Chapter 7 explain how network analysis enables decision-making systems to be visualised as well as analysed. They illustrate the use of network analysis by research into landscape governance in Maluku, Indonesia, and reflect on research design, data collection and analytical methods.

The following four chapters draw more on ethnographic methods to research plurality and hybridity of institutions, reflecting more of a critical institutionalism approach. In Chapter 8, Datta recounts how she used multiple methods to research local institutions in forest governance in India. In particular, she reflects on the benefits of researchers immersing themselves in the research situation to gain trust and generate a rich understanding of the nature and dynamics of institutions. Participant observation was found to be particularly important in enabling deep analysis of the institutional arrangements. Chapter 9 takes critical institutionalism further by considering non-human entities in water governance, reporting on the development and application of socio-technical analysis of institutions. Chitata, Cleaver and Kemerink-Seyoum demonstrate how the study of non-human entities such as irrigation infrastructure can generate insights into the functioning of institutions.

Chapters 10 and 11 draw on assemblage thinking to investigate institutional diversity. Orihuela in Chapter 11 introduces an assemblage theoretical framework and the use of actor network theory to research institutional diversity in the context of forest governance in Peru, using historical analysis and ethnographic fieldwork. In Chapter 11, Tsering and Unks also report on using ethnographic methods and draw on assemblage thinking to investigate the hybridity and fluidity of institutions in pastoralism in China and Kenya. Rationale for the use of multiple methods, including interviews, participant observation and photovoice, provides guidance on method selection, supported by reflections on practical and ethical concerns.

The concluding chapter identifies lessons and insights on the methods and reflections from across the preceding chapters, providing an overview of key considerations for researchers in planning for, and undertaking, research into institutions associated with natural resource governance.

References

Aggarwal, R.M. and Anderies, J.M. (2023) Understanding how governance emerges in social-ecological systems: Insights from archetype analysis, *Ecology and Society*, 28(2): 2. https://doi.org/10.5751/ES-14061-280202

Arts, B. (2014) Assessing forest governance from a 'Triple G' perspective: Government, governance, governmentality, *Forest Policy and Economics*, 49: 17–22. https://doi.org/10.1016/j.forpol.2014.05.008

Basurto, X., Kingsley, G., McQueen, K., Smith, M. and Weible, C.M. (2010) A systematic approach to institutional analysis: Applying Crawford and Ostrom's Grammar, *Political Research Quarterly*, 63(3): 523–537. https://doi.org/10.1177/1065912909334430

Campese, J. (2016) Natural Resource Governance Framework Assessment Guide: Learning for improved natural resource governance, *IUCN/CEESP NRGF Working Paper*, Gland: IUCN and CEESP.

Carlisle, K. and Gruby, R.L. (2019) Polycentric systems of governance: A theoretical model for the commons, *Policy Studies Journal*, 47(4): 927–952. https://doi-org.bham-ezproxy.idm.oclc.org/10.1111/psj.12212

Cleaver, F. (2002) Reinventing institutions: Bricolage and the social embeddedness of natural resource management, *European Journal of Development Research*, 14(2): 11–30. https://doi.org/10.1080/714000425

Cleaver, F. (2012) *Development through Bricolage: Rethinking Institutions for Natural Resource Management*, London: Earthscan.

Cleaver, F. and Whaley, L. (2018) Understanding process, power, and meaning in adaptive governance: A critical institutional reading, *Ecology and Society*, 23(2): 49. https://doi.org/10.5751/ES-10212-230249

Colding, J. and Folke, C. (2001) Social taboos: "Invisible" systems of local resource management and biological conservation, *Ecological Applications*, 11(2): 584–600. https://doi.org/10.1890/1051-0761(2001)011[0584:STISOL]2.0.CO;2

Cox, M., Arnold, G. and Villamayor Tomás, S. (2010) A review of design principles for community-based natural resource management, *Ecology and Society*, 15(4): 38. https://www.ecologyandsociety.org/vol15/iss4/art38/

Crawford, S.E.S. and Ostrom, E. (1995) A grammar of institutions, *The American Political Science Review*, 89(3): 582–600. https://www.jstor.org/stable/2082975

Crona, B., Ernstson, H., Prell, C., Reed, M. and Hubacek, K. (2011) Combining social network approaches with social theories to improve understanding of resource governance, in Bodin, Ö and Prell, C. (eds.) *Social Networks and Natural Resource Management: Uncovering the Social Fabric in Environmental Governance*, Cambridge: Cambridge University Press.pp. 44–71.

de Koning, J. (2014) Unpredictable outcomes in forestry-governance institutions in practice, *Society and Natural Resources*, 27(4): 358–371. https://doi.org/10.1080/08941920.2013.861557

de Koning, J. and Cleaver, F. (2012) Institutional bricolage in community forestry: An agenda for future research, in Arts, B., van Bommel, S., Ros-Tonen, M. and Verschoor, G. (eds.) *Forest-People Interfaces*, Wageningen: Wageningen Academic Publishers, pp. 277–290.

Epstein, G., Morrison, T.H., Lien, A., Gurney, G.G., Cole, D.H., Delaroche, M., Villamayor Tomas, S., Ban, N. and Cox, M. (2020) Advances in understanding the evolution of institutions in complex social-ecological systems, *Current Opinion in Environmental Sustainability*, 44: 58–66. https://doi.org/10.1016/j.cosust.2020.06.002

Epstein, G., Villamayor-Tomas, S. and Schoon, M. (2021) Institutional analysis, in Biggs, R., de Vos, A., Preiser, R., Clements, H., Maciejewski, K. and Schlüter, M. (eds.) *The Routledge Handbook of Research Methods for Social-Ecological Systems*, London: Routledge, pp. 307–320.

Funder, M. and Marani, M. (2015) Local bureaucrats as bricoleurs. The everyday implementation practices of county environment officers in rural Kenya, *International Journal of the Commons*, 9(1): 87–106. https://doi.org/10.18352/ijc.526

Haapala, J., Rautanen, S.-L., White, P., Keskinen, M. and Varis, O. (2016) Facilitating bricolage through more organic institutional designs? The case of water users' associations in rural Nepal, *International Journal of the Commons*, 10(2): 1172–1201. https://doi.org/10.18352/ijc.688

Hall, K., Cleaver, F., Franks, T. and Maganga, F. (2014) Capturing critical institutionalism: A synthesis of key themes and debates, *European Journal of Development Research*, 26(1): 71–86. https://doi.org/10.1057/ejdr.2013.48

Helmke, G. and Levitsky, S. (2004) Informal institutions and comparative politics: A research agenda, *Perspectives on Politics*, 2(4): 725–740. https://www.cambridge.org/core/journals/perspectives-on-politics/article/abs/informal-institutions-and-comparative-politics-a-research-agenda/11A6AA9DCAF579F5B72DA250728AA513#

Jupille, J. and Caporaso, J.A. (2022) *Theories of Institutions*, Cambridge: Cambridge University Press.

Kairu, A., Upton, C., Huxham, M., Kotut, K., Mbeche, R. and Kairo, J. (2018) From shiny shoes to muddy reality: Understanding how meso-state actors negotiate the implementation gap in participatory forest management, *Society and Natural Resources*, 31(1): 74–88. https://doi.org/10.1080/08941920.2017.1382628

Karambiri, M., Brockhaus, M., Sehring, J. and Degrande, A. (2020) 'We are not bad people'- Bricolage and the rise of community forest institutions in Burkina Faso, *International Journal of the Commons*, 14(1): 525–538. https://doi.org/10.5334/ijc.1061

Labrousse, A. (2016) Not by technique alone. A methodological comparison of development analysis with Esther Duflo and Elinor Ostrom, Journal of Institutional Economics, 12(2): 277–303. doi:10.1017/S1744137415000429

Lauth, H.-J. (2000) Informal institutions and democracy, *Democratization*, 7(4): 21–50. https://doi.org/10.1080/13510340008403683

Lien, A.M. (2020) The institutional grammar tool in policy analysis and applications to resilience and robustness research, *Current Opinion in Environmental Sustainability*, 44: 1–5. https://doi.org/10.1016/j.cosust.2020.02.004

Lowndes, V. and Roberts, M. (2013) *Why Institutions Matter: The New Institutionalism in Political Science*, Basingstoke: Palgrave Macmillan.

Mahoney, J. and Thelen, K. (2010) A theory of gradual institutional change, in Mahoney, J. and Thelen, K. (eds.) *Explaining Institutional Change: Ambiguity, Agency, and Power*, New York: Cambridge University Press, pp. 1–37.

Marin, A. and Bjørklund, I. (2015) A tragedy of errors? Institutional dynamics and land tenure in Finnmark, Norway, *International Journal of the Commons*, 9(1): 19–40. https://doi.org/10.18352/ijc.586

McGinnis, M.D. (2011) An introduction to IAD and the language of the Ostrom workshop: A simple guide to a complex framework, *Policy Studies Journal*, 39(1): 169–183. https://doi.org/10.1111/j.1541-0072.2010.00401.x

North, D.C. (1990) *Institutions, Institutional Change and Economic Performance*, Cambridge: Cambridge University Press.

Nunan, F. (2020) Conclusion: Working towards diverse, flexible and inclusive natural resource governance, in Nunan, F. (ed.) *Governing Renewable Natural Resources: Theories and Frameworks*, Abingdon: Routledge, pp. 226–244.

Osei-Tutu, P., Pregernig, M. and Pokorny, B. (2015) Interactions between formal and informal institutions in community, private and state forest contexts in Ghana, *Forest Policy and Economics*, 54: 26–35. https://doi.org/10.1016/j.forpol.2015.01.006

Ostrom, E. (1990) *Governing the Commons: The Evolution of Institutions for Collective Action*, Cambridge: Cambridge University Press.

Ostrom, E. (2005) *Understanding Institutional Diversity*, Princeton, NJ: Princeton University Press.

Ostrom, E. (2007) A diagnostic approach for going beyond panaceas, *Proceedings of the National Academy of Sciences*, 104: 15181–15187. https://doi.org/10.1073/pnas.0702288104

Ostrom, E. (2009) A general framework for analyzing sustainability of social-ecological systems, *Science*, 325: 419–422. https://doi.org/10.1126/science.1172133

Ostrom, E. (2011) Background on the institutional analysis and development framework, *Policy Studies Journal*, 39(1): 7–27. https://doi.org/10.1111/j.1541-0072.2010.00394.x

Ostrom, E., Dietz, T., Dolsak, N., Stern, P.C., Stonich, S. and Weber, E.U. (eds) (2002) *The Drama of the Commons*, Washington, DC: National Academies Press.

Partelow, S. (2018) A review of the social-ecological systems framework: Applications, methods, modifications, and challenges, *Ecology and Society*, 23(4): 36. https://doi.org/10.5751/ES-10594-230436

Po, J.Y.T., Saint Ville, A.S., Rahman, H.M.T. and Hickey, G.M. (2019) On institutional diversity and interplay in natural resource governance, *Society & Natural Resources*, 32(12): 1333–1343. https://doi.org/10.1080/08941920.2019.1667463

Poteete, A.R., Janssen, M.A. and Ostrom, E. (2010) *Working Together: Collective Action, the Commons, and Multiple Methods in Practice*, Princeton: Princeton University Press.

Rahman, H.M.T., Saint Ville, A.S., Song, A.M., Po, J.Y.T., Berthet, E., Brammer, J.R., Brunet, N.D., Jayaprakash, L.G., Lowitt, K.N., Rastogi, A., Reed, G. and Hickey, G.M. (2017) A framework for analyzing institutional gaps in natural resource governance, *International Journal of the Commons*, 11(2): 823–853. https://doi.org/10.18352/ijc.758

Rival, L. and Muradian, R. (2013) Introduction: Governing the provision of ecosystem services, in Muradian, R. and Rival, L. (eds.) *Governing the Provision of Ecosystem Services*, Dordrecht: Springer, pp. 1–17.

Sandström, C. (2009) Institutional dimensions of comanagement: Participation, power, and process, *Society & Natural Resources*, 22(3): 230–244. https://doi.org/10.1080/08941920802183354

Schlüter, A. and Theesfeld, I. (2010) The grammar of institutions: The challenge of distinguishing between strategies, norms, and rules, *Rationality and Society*, 22(4): 445–475. https://doi.org/10.1177/1043463110377299

Tucker, C.M., Hribar, M.S., Urbanc, M., Bogataj, N., Gunya, A., Rodela, R., Sigura, M. and Piani, L. (2023) Governance of interdependent ecosystem services and common-pool resources, *Land Use Policy*, 127: 106575. https://doi.org/10.1016/j.landusepol.2023.106575

Whaley, L. (2018) The Critical Institutional Analysis and Development (CIAD) framework, *International Journal of the Commons*, 12(2): 137–161. https://doi.org/10.18352/ijc.848

Yeboah-Assiamah, E., Muller, K. and Domfeh, K.A. (2017) Institutional assessment in natural resource governance: A conceptual overview, *Forest Policy and Economics*, 74: 1–12. https://doi.org/10.1016/j.forpol.2016.10.006

Young, O.R. (2002) *The Institutional Dimensions of Environmental Change: Fit, Interplay, and Scale*, Cambridge: The MIT Press.

2 Using the Institutional Analysis and Development (IAD) framework to link collective action, behavioral change and social-environmental outcomes in conservation policy

Tanya Hayes and Felipe Murtinho

Introduction

Over the past decades, practitioners and policymakers have implemented a range of decentralized conservation policies and programs to promote natural resource management and livelihood development in impoverished, resource-dependent communities (Andersson, Gibson and Lehoucq, 2006; Larson and Soto, 2008; Meinzen-Dick, 2007; Ribot, 2003). A challenge we face is understanding if and how said initiatives attain the desired resource use behaviors and associated outcomes (Ferraro and Pattanayak, 2006; Pynegar et al., 2021). This is particularly true in community-based conservation programs and similar incentive-based programs implemented in the communal context where programs often depend on communal governance mechanisms to transmit the program policies, and associated benefits, to the broader community (Agrawal and Gibson, 1999; Hayes, Murtinho and Wolff, 2015; Kerr, Vardhan and Jindal, 2014).

The Institutional Analysis and Development (IAD) framework developed by Elinor Ostrom and colleagues, with its focus on collective action, provides a systematic means to understand how institutions interact with local biophysical and socioeconomic conditions to influence communal resource management processes and outcomes (Ostrom, 1990, 2005). Specifically, in considering how conservation policies and programs influence local resource management decisions, we have found the IAD framework particularly helpful in teasing out how 'formal' institutions, such as protected area policies or payment for ecosystem services (PES) contracts, interact with community institutions, or the rules and norms that are recognized, and shape behavior, in a specific locale (Cole, 2017; Ostrom, 2011).

Thus, the purpose of this chapter is to illustrate how the IAD framework can be applied to assess how decentralized conservation policies implemented in communal settings are transmitted via collective decision-making processes into locally understood rules, which in turn shape household conservation behaviors and the respective social and environmental outcomes. To do so, we use examples from our research on a payment for ecosystems services (PES)

DOI: 10.4324/9781032624679-2

program in Ecuador to walk through how we applied the IAD framework and associated methods to assess institutional development and behavioral change in rural communities in the Ecuadorian highlands. The examples highlight the strengths of the IAD framework as well as some of the decisions and challenges that a researcher must address when implementing the study in 'real-world' contexts.

In the following, we provide a brief context to understand our study in Ecuador. We then lay out some of the key components of the IAD framework and how we applied the framework to our research design, data gathering, and analysis. In discussing our application of the framework to 'real world' resource management situations, we highlight strategies and challenges of (i) using IAD to craft comparative case study research designs to assess the outcomes of different resource management policies; (ii) measuring institutional change and the interaction between formal conservation policies and communal governance arrangements; (iii) identifying and measuring land-use behaviors and behavioral change; and (iv) triangulating quantitative and qualitative data at the community and household levels to empirically link program intervention to institutional and behavioral change in support of specific conservation outcomes.

Study context

The examples used in this chapter come from our study on community resource management in the Ecuadorian highlands, a South American mountain range that runs north-south through the center of Ecuador. It is a densely populated region where indigenous and mestizo communities (of mixed Spanish and indigenous descent) live alongside ecologically valuable high-montane ecosystems. These communities depend on the land for subsistence and semi-subsistence activities, including potato production and the grazing of sheep and cattle (Bremer et al., 2014; Buytaert et al., 2006).

In 2009, as part of their national PES program, the Ecuadorian government created a sector specifically to protect the páramo, a high-elevation ecosystem of grasslands and shrubs (at about 3,500 m) that provides valuable ecosystem services including water regulation, biodiversity protection, and carbon sequestration. As part of the program, rural communities signed collective conservation contracts with the Ecuadorian government in which they agreed to conserve their highlands ecosystems (i.e., reduce grazing and prohibit agriculture and hunting) for a 20-year period. In return, communities received collective payments twice yearly on the condition that they prohibit burning and restrict resource use (grazing, agriculture, and hunting) on their communal lands (Hayes et al., 2022).

Our work intentionally focused on collective PES contracts with communities as communal PES programs present governance challenges that are distinct from the individual PES model. In the individual model, a landowner signs a PES contract and agrees to receive payment on the condition that they comply with specific resource use conditions. In collective contracts, however, a communal governing body must facilitate the decision to join, transmit the desired behavioral changes to the rest of the community, and decide how payments will be distributed. Thus, unlike individual PES contracts, in collective PES, the communal decision-making processes can be critical in influencing program outcomes (Hayes, Murtinho and Wolff, 2015; Kerr, Vardhan and Jindal, 2014).

The implemention of collective PES contracts raises concerns about the ability of communities to craft rules that produce the desired behavioral changes and result in additional conservation outcomes (Dougill et al., 2012; Kerr, Vardhan and Jindal, 2014; Kosoy, Corbera and Brown, 2008; Muradian, 2013). Likewise, research in community-based natural resource management points to the potential for elite capture of program benefits and questions the ability of communities to equitably distribute the costs and benefits derived from participation across all households (Agrawal and Gibson, 1999; Cleaver, 2017; Cooke and Kothari, 2001; Hayes et al., 2019; Pascual et al., 2014).

Our study sought to address the above-mentioned concerns by examining how particpation in the Ecuadorian PES program influence community land-use rules, land-use behaviors, and equity concerns regarding the decision to participate and the distributions of costs and benefits across households. Our study began in 2012 and continued through 2022. We used the IAD framework to craft a general conceptual framework and related models of the potential factors that could influence the impact of the PES program on land-use behaviors and additional governance outcomes. This, in turn, shaped our research design, data gathering, and analysis. In this chapter, we discuss how we used the IAD framework to examine: (i) how participation in the PES program influenced the creation of land-management rules, and (ii) the program's impact on household land-use decisions.[1]

Application of the IAD framework to assess the links between collective action, behavioral change, and program outcomes

As discussed in Chapter 1, the IAD framework provides researchers with a structured set of categories to map out how institutions, in relation to other contextual conditions and attributes of the actors involved, shape interactions, decisions, and behavioral outcomes (Ostrom, 2005). At the core of the IAD framework is the *action situation*; the social spaces where individuals and organizations interact to exchange goods, negotiate, and act collectively, engaging in discussions, making rules, and mediating conflicts (among other interactions) (Ostrom, 2011).

In its most general form, the framework highlights a set of biophysical, community, and institutional variables that may influence an action situation and related outcomes of interest. Within the IAD, however, there are multiple ways to further unpack the analysis, including, but not limited to digging into the rule arrangements that shape an action situation, identifying specific contextual factors or attributes of the actors involved, and exploring the interactions across multiple action situations (Ostrom, 2005, 2011). Scholars have also explored potential ways to combine IAD with the related social-ecological systems (SESs) framework to provide a more refined analysis of complex SESs (Cole, Epstein and McGinnis, 2019; Ostrom, 2011; Ostrom and Cox, 2010). As Ostrom notes, the framework itself does not prescribe any specific theories. Rather, it provides a systematic means to identify variables and conceptualize processes that, in turn, can be used to craft models, and test predictions about factors influencing an outcome of interest (Ostrom, 2011).

In our study, we used the IAD framework to create a conceptual map to, in turn, develop theoretically grounded models and associated hypotheses to understand the influence of the PES program on a set of communal governance institutions and household behavioral outcomes. To examine the role of institutions, we differentiated between formal legal institutions, in this case the PES contract rules, and the communities' own 'rules-in-use', locally recognized institutions that shaped community members' land-use and management decisions (Cole, 2017; Ostrom, 2011; see Chapter 1 for further explanation of formal institutions and local or informal institutions).

In environmental conservation, Ostrom and others have noted the problem of formal institutions, such as protected area policies or environmental laws, remaining on paper only; failing to influence decisions on-the-ground (Cole, 2017; Gibson, McKean and Ostrom, 2000; Ostrom and Nagendra, 2006; Ostrom, 2011). At the same time, institutional researchers and others have demonstrated the wealth of locally recognized rules and norms that often shape day-to-day resource use (Cole, 2017; Dietz, Ostrom and Stern, 2003; Gibson, McKean and Ostrom, 2000). Our study sought to assess if and how communal participation in the PES program, specifically the conservation conditions (formal rules) of the PES contracts, influenced community land management institutions, namely the 'rules-in-use', and household land-use decisions to support the conservation of the respective community's collective páramo (Ostrom, 2011).

Figure 2.1 shows how we applied the IAD framework to understand land management institutions and land-use behaviors in our study sites. Note, we slightly modified the IAD framework to consider how a set of communal governance conditions influence household land-use behavior (aligning the IAD framework with the SES framework), and to dig further into how specific household attributes influenced PES outcomes. We also drew from the SES framework to consider PES outcomes given the broader socio-political, economic, and related ecosystem characteristics, namely climate change (Ostrom, 2009).

In our study, the focal action situations were communal assembly meetings where community members made collective decisions about the use of collective lands, as well as the lands where they grazed animals, hunted, and engaged in other resource-related activities. We identified two related outcomes from these action situations: (i) the creation of new land-use rules (initial output)

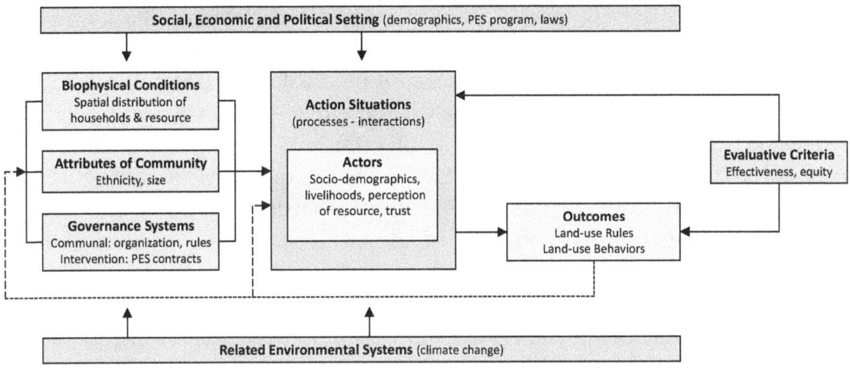

Figure 2.1 IAD framework adapted to payment for ecosystems services (PES) assessment.

and (ii) land-use behaviors (longer term outcome). We hypothesized that in deciding to participate in the PES program, a community would create new land-use rules and, in turn, households would engage in new land conservation behaviors. We also hypothesized, however, that land-use behaviors do not operate in a vacuum, recognizing the potential influence of household attributes, local biophysical conditions, attributes of the communities and the respective governance systems, and the broader social, economic, political, and environmental setting.

With respect to household attributes, in our study, we consider actors to be boundedly rational and to make decisions based on the control they perceive to have over that decision, the information they have regarding their choice options, the likely outcomes, and the benefits and costs (broadly defined by a set of monetary and non-monetary values) they perceive for these outcomes (Jones, 1999; Ostrom, 2005). Drawing on previous work on the social and cognitive factors that influence natural resource use and behavior, we considered how socioeconomic and demographic variables, such as age of household head, education, gender, livelihoods, and cognitive factors such as perceived value of the resource, and trust in community and government programs, influenced participation in communal decision-making and land-use behaviors (Beedell and Rehman, 2000; Grothmann and Patt, 2005; Koontz, 2001; Vatn, 2010; Vignola et al., 2010).

We also considered the local contextual conditions often predicted to influence local environmental governance and resource use decisions (Ostrom, 2009). In our research design, we tried to keep the biophysical conditions largely consistent across communities; we accounted for differences in páramo size, and the population density and spatial distribution of the households (proximity to páramo and town center). In the IAD, community attributes often include shared values or beliefs, in addition to socioeconomic and demographic variables. In our study, many of those variables were measured at the household level. We did, however, account for the ethnicity of the community and the size (Poteete and Ostrom, 2004).

Of particular interest in our study were the attributes of the communal governance systems. Our principal independent variable for the study was whether the community had decided to join the PES program and thereby signed the PES contract (formal policy intervention). We were, however, also interested in understanding the role of communal governance conditions in shaping household land-use behavior. Based on previous literature, we hypothesized that existent community rules regarding land-use, monitoring and enforcement, and decision-making, as well as the overall organization and collective decision-making capacity of the respective communities, would also shape land-use behavior and, potentially, the ability of PES to attain its desired conservation behaviors in the participant communities (Andersson, Benavides and León, 2014; Gibson, Williams and Ostrom, 2003; Muradian, 2013).

Lastly, we did not include the broader scale socioeconomic, political, and climatic conditions (apart from the PES program) in our statistical models as they were relatively constant across communities. We did, however, examine these conditions to better contextualize our results to consider how regional economic and demographic shifts, as well as climatic changes may have shaped the outcomes of our study.

From framework to implementation: research design, data gathering, and analysis

The beauty of the IAD framework (and related SES framework) is its ability to capture numerous social, economic, political, biophysical, and cognitive factors that a researcher might want to consider in diagnosing and/or predicting a resource management outcome. The challenge, however, is how to move from an often-complex conceptual map to a concrete research design that will enable us to gather specific pieces of data to answer a defined research question and, as appropriate, test-related hypotheses. In the following, we discuss how we addressed these challenges in our research design, data gathering, and analysis.

Research design to evaluate an existing program

To answer our research question about the impacts of PES, we struggled with how to design a study that would provide us the breadth, depth, and analytical rigor to assess the influence of PES on land-use governance and behavior, while also taking into the account the practical programmatic and geographic realities of our study. In an ideal world, to measure program impact, we would have worked with the PES program personnel prior to the start of the program to develop a randomized control trial and gathered baseline data on program outcomes of interest before rolling out the program (Ferraro and Hanauer, 2014; Pynegar et al., 2021). Our study, however, began in 2012, roughly three years after the PES program had entered the region. Furthermore, communal participation in the program was voluntary and happened over time based on community decision-processes and program resources. At the time of our study, there were 47 communities participating in the program. Moreover, to answer our research questions we needed a research design that would enable us to examine not only if the PES program impacted behavior but also how and why (Pynegar et al., 2021).

Given these study limitations and our research questions, we decided to use a two-stage approach in our research design and data gathering. Stage I sought to provide a breadth of understanding on land management and land-use across communities in the highlands, and their participation in the PES program. Stage II used a more rigorous research design with a smaller subset of communities to determine the influence of PES participation on household land-use behavior.

In Stage I, we conducted structured interviews with community leaders in almost all participating communities (n = 44 communities), and in a set of comparison communities. The comparison communities were neighboring communities that collectively owned páramo lands in the program target areas and were, therefore, eligble to participate in the PES program, but had not decided to participate (n = 23 communities). Due to the potential inherent selection bias in the decision to participate in PES, the comparison communities were limited in their ability to act as a control group to assess the impact of PES participation. Nonetheless, interviews with former and current leaders served to identify the factors associated with the decision to participate, and to trace the history of páramo use and associated land-use rules across the highlands, and identify rule changes that occurred after particpating communities joined the PES program. This data also informed our selection of communities for Stage II of our study.

In Stage II, we used a quasi-experimental research design to assess how PES participation influenced household behaviors in select subset of communities (n = 12 communities) (Shadish, Cook and Campbell, 2002). In selecting communities, we sought to address the problem of potential selection bias in the decision to join the PES program by taking advantage of the gradual roll-out of the PES program in the highland communities. We purposefully matched a set of participating communities (n = 6) to a set of communities that were waiting to enroll in the program but had not yet had the opportunity to do so (our 'control' communities) (n = 6). The IAD framework informed our matching criteria to identify household and contextual factors that could be observable sources of bias. Communities were selected based on the following criteria: identify with the same indigenous ethnicity (Quichua); households depend principally on farm-level activities for their livelihoods; communities had been using the páramo prior to 2008 when the PES program entered the region; the páramo was located at relatively similar altitudes and with similar topography; and, most residents could access the páramo by walking from their houses in less than three hours. The communities were representative of the poorer parishes in the region and of the distribution of community sizes and páramo sizes common to the region.[2]

We gathered data over roughly two periods of time (2013 and 2018). To assess the impacts of PES on land-use, within each community, we conducted household surveys (*n*=420 in 2013, and *n*=472 in 2018), combined with observed land-use assessments, land cover satellite images, focus group discussions, and leader interviews. Our analyses included qualitative and quantitative methods, including logit regresssion models and the difference-in-difference (DID) approach to assess program impact. We provide further explanation of our data gathering and analyses in the subsequent sections.

Institutional change: identifying and measuring[3]

Central to our study was the ability to define relevant land management rules and identify communal collective action that resulted in rule changes. The PES contracts (formal institutions) prohibited burning, hunting, practicing agriculture, or introducing non-native species or any activities that could impact the conservation value of the designated area. The contract also stipulated that communities limit grazing in the páramo to less than semi-intensive levels, although no formal guidelines were given as to what constitutes semi-intensive levels (Ministerio del Ambiente, 2009). In the highlands, after signing the PES contract, communities were then responsible for crafting appropriate management rules and applying those rules (creating their own 'rules-in-use'). While PES program personnel were responsible for overseeing compliance, communities were responsible for monitoring páramo use and reporting any transgressions to PES program personnel (Hayes, Murtinho and Wolff, 2015).

As shown in Figure 2.2, we considered three points to identify the presence and application of land-use rules in the communities: (i) rule presence and specification (whether an activity was prohibited or restricted), (ii) monitoring, and (iii) enforcement. To examine the presence of land-use rules and rule change, we considered whether communities had rules related to access and use of the páramo for specific activities such as grazing, hunting, or agriculture.[4] We asked whether different activities were permitted, prohibited, or restricted based on certain conditions such as timing, place, quantity, or attributes of the household (Crawford and Ostrom, 1995).

Designating a rule does not necessarily indicate that the rule is recognized and applied (Ostrom, 1990, 2011). To assess the degree to which the rule was applied, we drew from work on the grammar of institutions syntax that helps to distinguish between communal norms and strategies, and 'rules-in-use'. Rules-in-use, in contrast to norms or strategies, are monitored and enforced with consequences for failure to comply (Crawford and Ostrom, 1995; Ostrom, 2005). In our study, we asked whether the community had an organized system to monitor compliance with páramo rules. In those communities that were organized, we asked whether their monitoring systems were based on the rotation of community households, or whether they hired a local to work as a forest guard (often a stricter monitoring mechanism). And, finally, to enforce land-use rules, we considered whether there were any sanctioning

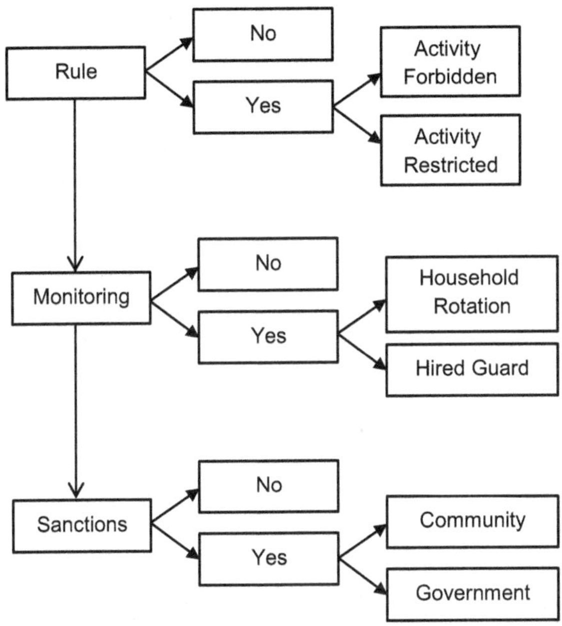

Figure 2.2 Institutional arrangements and points for rule change (Hayes, Murtinho and Wolff, 2015: 83).

mechanisms that were clearly defined, and if so, whether the community had its own sanctioning system, or whether they relied on external actors, such as the government, to apply sanctions for rule transgressions.

Rule presence, monitoring, and enforcement mechanisms served to identify the current institutional arrangements for land management and pointed to areas where communal collective decision-making processes may respond to participation in PES by creating and/or strengthening existing institutions. For example, at the first tier, a community may create a new rule or increase the restrictiveness of a present rule. In our study, we found that while some communities completely forbid grazing, many initially restricted grazing by limiting the number of animals a household may have in the páramo or designating specific areas for grazing. After participating in PES, however, a number of communities strengthened their grazing rules by moving from grazing allowed with restrictions to completely prohibiting grazing (Hayes, Murtinho and Wolff, 2015). Likewise, to apply a new or existing rule, a community may strengthen its organizational capacity to monitor its land-use rules. For example, we found that some PES participating communities moved from a more informal household rotation system for monitoring the páramo to paying people to serve as guards to monitor uses on their collective lands (Hayes, Murtinho and Wolff, 2015).

Data on communal land-management institutions was gathered in Stage I of our study via a questionnaire administered to community leaders in 44 participating in PES program and leaders in 23 non-participant communities. The questionnaire included a set of open-ended questions that drew from the International Forestry Resources and Institutions (IFRI) protocols (http://ifri.forgov.org) that we adapted to correspond to páramo land uses and our interest in PES. Leaders were asked about the history of rule creation (if rules were present) and their monitoring and enforcement mechanisms, if there were events or organizations that sparked the creation of rules or strengthened their applications, how they monitored the rules, challenges in compliance, and if and how sanctions were applied for non-compliance. For those that were participating in the PES program, leaders were then asked to recall the decision to participate in the PES program and if the rules or monitoring and enforcement mechanisms had changed because of participating in PES.

We originally intended to analyze all the different conditions in the institutional statements that limited resource use (e.g., who can graze, how many animals can be grazed, and/or if there are times or areas where grazing can occur). This, however, proved difficult. In addition to being very time consuming to record and code consistently, our statistical analyses could not capture all the rule variations. Thus, we ultimately limited our coding of rules to whether an activity was permitted, forbidden, or had some sort of restrictions. For those participating in PES, we also coded whether there was a change in rule specification, in the monitoring systems or in enforcement and sanctions. Although we asked about a variety of páramo uses (grazing, hunting, gathering grasses, etc.), we largely focused our analysis on rules for grazing. Grazing was the greatest use of the páramo in the study region, the area where we saw the greatest institutional variability across communities, and where the PES contract stipulated that communities were responsible for developing their own restrictions.

Behavioral change: identifying and measuring land-use behavior

To understand how PES influenced conservation of the collective lands, we chose to focus on the most common land-use behavior, namely whether households continued to graze their animals (cows and sheep) in the conservation lands. To gather data on household grazing behavior and trace it over time, we used a household questionnaire that we triangulated with other qualitative and biophysical data sources gathered in Stage II. We administered the household questionnaire in our participant and control communities in two points in time 2013, approximately four years into the program ($n = 420$ households) and again in 2018, after a period of program disruption ($n = 472$ households) (Hayes et al., 2022). Survey questions asked households if they had animals, where they grazed them during different times of the year, if they grazed in the collective páramo, and how grazing behavior had changed over time. To establish grazing prior to program participation, we asked about páramo use in 2008 (prior to the PES program entering the region) in the 2013 survey (Hayes, Murtinho and Wolff, 2017).

We used several tools to increase the validity of our questionnaire and the responses about current páramo use and past uses (Hayes, Murtinho and Wolff, 2017). First, prior to administering the questionnaire, we held focus group discussions with elders and leaders from different community organizations. In these discussions, participants drew a map of the community and páramo lands, and discussed primary land-uses and how those uses change over time, as well as governance issues, including the development of páramo use rules and compliance, on the communities' lands. Participants also created a timeline to establish key events and identify an event that occurred approximately five years earlier (circa 2008 prior to when the PES program entered the region). We used the focus group discussions to shape our questions about land-use practices and particularly, responses to seasonal changes. For the recall questions, interviewers used the timeline events created in the focus groups to establish a reference event close to 2008 (prior to PES entering the region) for each community (Raphael, 1987; Schaeffer and Presser, 2003).

Second, in structuring the questionnaire, questions positioned páramo use as a common livelihood strategy and gave respondents both closed and open-ended opportunities to discuss if and how they use the páramo and how their use had changed (Schaeffer and Presser, 2003; Thornberry and Krohn, 2000; Tourangeau and Yan, 2007). Furthermore, rather than ask about the number of animals a household grazed in the collective páramo and how many they grazed five years ago (information that can be sensitive and that would be very difficult to recall precisely), we asked about grazing behavior in terms of gradients of use (more today or less today) as this would be easier to recall. Later, we triangulated the gradients of use with other land use responses (from close and open-ended questions) to create the binary variable for grazing in 2008, 2013, and 2018 (Hayes, Murtinho and Wolff, 2017; Hayes et al., 2022).

Lastly, to further validate our questionnaire results, we triangulated the household information with a rapid field assessment of the páramo and analyses of aerial images. A rapid field assessment was conducted by a local environmental scientist in 2013 and 2018 to identify current land-uses in the páramo, the state of degradation, and evidence of cattle, sheep, and fires along a set of transect lines (Hayes, Murtinho and Wolff, 2017; Hayes et al., 2022). This data was further supplemented by a land use/land cover visual analysis of satellite images including Pleiades 1A (February 2019) and SPOT 6 and 7 (December 2017). This aerial analysis established the limits of the agricultural frontier in each community and identified if there was agriculture and grazing activities in the areas under the PES conservation contracts (Hayes et al., 2022).

Assessing program impacts: linking a program to institutional and behavioral change

Returning to the initial purpose of our study and the conceptual models derived from our IAD framework, in analyzing our data we sought to answer three inter-related questions about the influence of PES on land-management

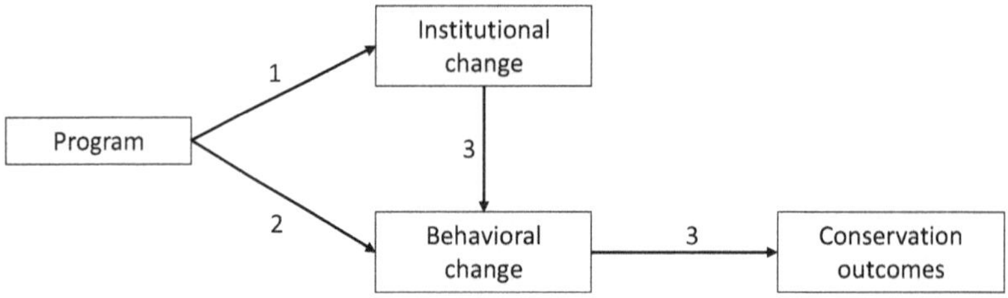

Figure 2.3 Steps to link the program to institutions, behavior, and conservation.

and conservation of the communal lands. Figure 2.3 lays out the analytical steps we took to answer our questions; namely, to assess how the program influenced institutional change (Step 1), how the program influenced behavioral change (Step 2), and ultimately, to assess the link between program, institutional change, behavioral change, and the ability to attain the desired conservation outcomes (Step 3). Our analyses drew on data gathered in Stages I and II (as explained above). In the following, we discuss the methods we used, as well as some of the challenges we faced, in answering our questions per each analytical step.

Linking a program to institutional changes

As shown in Figure 2.3, the first question we sought to understand was how participation in the PES program (and associated contract conditions) influenced communal collective action in creating new land management institutions (rules) (Step 1). To analyze rule development and specifically trace if and how the PES program influenced the development of land-use rules and their applications, we used descriptive statistics coupled with qualitative data on the history of rule development in the communities. As discussed earlier, given the research design limitations in our sample of highland communities (44 participating and 23 non-participating) in Stage I of our study, our analysis of institutional change was limited to the use of bivariate statistics combined with findings from our open-ended questionnaire that asked leaders about the history of rule development to demonstrate an association between PES participation and the presence of stricter land-use rules, although it was less clear the degree to which those rules were in fact, applied.[5]

To further understand rule application within the communities, we complemented the leader survey data gathered in Stage I with data from the household questionnaire and focus group discussions on rule history from Stage II. The survey served to assess how participant households perceived the influence of PES participation on land-use rules and compared perceived rule monitoring and enforcement across all households (participant and non-participant) by asking the respondent whether they considered it likely that a rule-breaker will be caught, and if they know of a rule-breaker that was caught in the previous year.

We found that while all communities had developed at least some land-use regulations over the past ten years, PES communities were significantly more likely to prohibit grazing on their collective lands than non-participant communities. When we traced rule development in the communities, we found that PES communities were likely to have strengthened their land-use rules since participation, often moving rules that restricted grazing to those that now completely prohibited grazing in the collective lands. It was questionable, however, the degree to which the rules were applied (Hayes, Murtinho and Wolff, 2015). Our findings further illustrated the importance of disaggregating institutional arrangements to parse the strength of the statement (permits, forbids, conditional restrictions), and to assess the monitoring and enforcement mechanisms created to apply a rule.

Linking a program to behavioral changes in the context of local social-environmental conditions

The second question we sought to answer was the impact of participating in PES on household land-use behaviors, while recognizing how local biophysical, communal governance, and household factors identified in our IAD framework might also influence land-use decisions (Figure 2.3, Step 2). To test for the impact of PES on land-use behavior (in this case, the household decision to graze animals in the collective páramo), we performed a series of DID regressions, controlling for the household and contextual characteristics also predicted to impact grazing. The DID results found that PES produced an almost 20% reduction in the average number of households grazing animals in the PES communities, and that the impact remained robust, even when considering other household and contextual factors (Hayes, Murtinho and Wolff, 2017; Hayes et al., 2022).[6]

The results also indicated, however, that in addition to PES, household and governance conditions also influenced grazing behaviors. Thus, to complement the DID analysis, we ran a set of logit models that tested for how communal land-use rules and household demographic, socioeconomic and cognitive factors on land-use decisions, in addition to participation in PES, influenced the decision to reduce grazing on the collective lands. The logit analyses enabled us to identify how local institutions, specifically a pre-existing history of creating land-use rules, and the perceived application of monitoring and enforcement mechanisms decreased grazing by households across all communities (participant and non-participant) (Hayes, Murtinho and Wolff, 2017).

Linking a program, the institutional and behavioral changes, and the conservation outcomes

Lastly, our study sought to understand the links between PES, communal governance conditions, households, and their respective land-use decisions, and ultimately the conservation outcomes (Figure 2.3, Step 3). To tease out the link between PES, land-use rule change, and behavioral change, we used survey data, focus group and key informant interviews, and observational data to compare institutions, land-use behaviors, and land-conservation across the seven PES participating communities that were part of Stage II in our study.[7]

Bivariate analysis of the association between a community's organizational capacity, the changes made in its governance institutions after participating in PES, and changes in household grazing behavior illustrated the potential mediating role that communal governance capacity may play in implementing PES. The results also pointed to the potential for a reciprocal reinforcing relationship between PES and community organizations. Specifically, bivariate statistics and focus group findings indicated that PES performance varied depending on community governance conditions, namely those with a history of rulemaking and a strong organizational capacity attained greater reductions in grazing. Those communities that were more organized and with pre-existing land-use rules, however, were also more likely to have strengthened those rules and their applications upon joining PES, thereby increasing their organizational capacity, and in turn, producing greater change in household land use behaviors (Hayes, Murtinho and Wolff, 2017).

Observational data further confirmed that the institutional and behavioral changes were resulting in the desired conservation outcomes, with respect to land-use, from PES participation. We used field assessments and analyses of aerial images to compare land-uses in participant and non-participant communities, and the respective collective lands designated for conservation. Our field analysis found greater evidence of grazing and other extractive land-use activities in non-participant communities as compared to the participant. Likewise, our analyses of aerial images indicated continued conservation within the boundaries of the designated lands (Hayes, Murtinho and Wolff, 2017; Hayes et al., 2022).

We couched the findings within the broader socioeconomic, political, and environmental context. As shown in our application of the IAD framework, in addition to internal biophysical, communal, and local governance conditions that we measured in the respective communities, households in the region were also responding to broader socio, political and climatic changes in the area. Specifically, demographic and economic shifts and a changing climate were also making high altitude grazing less desirable. While these contextual factors do not negate the impacts of the PES program on land-uses in the region, the local contextual factors measured in our analysis, as well as the broader socioeconomic and environmental context, help to assess the external validity of the findings and their applicability of the case to other regions that may have differing communal contexts and may not be experiencing similar contextual changes (Hayes et al., 2022).

Reflection on application of IAD to assess institutional and behavioral change

In our research, we find ourselves continually returning to the IAD framework (and associated adaptations) to help shape our study questions and our research design. We have found that the framework enables us to disentangle complex socio-environmental situations, tease out key contextual variables, and identify the institutional arrangements that influence our outcomes of interest. It provides a holistic means to identify relationships between multiple social and environmental systems, thereby avoiding the pitfalls of looking at an institution in a vacuum.

For researchers wishing to understand the role of institutions in natural resource governance, the IAD framework is particularly strong in providing a systematic structure to identify and map out how policies and laws (formal institutions) interact with local resource-management rules and norms, and how these interactions shape resultant land-use behaviors. As discussed in this chapter, we find the work by Schlager and Ostrom (1992) and Crawford and Ostrom (1995) particularly helpful in identifying and defining all relevant institutions, and for identifying institutional change.

In our experience, however, moving the IAD framework from paper to the field is not without its challenges and limitations. For one, identifying institutions in the field can be difficult as locally recognized norms and rules are not always written or verbalized in a codified manner. In addition to the work by Schlager and Ostrom (1992) and Crawford and Ostrom (1995) to categorize the institutional statements, we recommend the IFRI protocols (http://ifri.forgov.org) as a practical fieldwork guide for gathering data on institutions, resource-use behaviors, and other variables that often influence resource-use decisions. While the IFRI protocols are written to understand forest management, they can be easily modified to understand a range of resource-use behaviors (Ostrom et al., 1989).

Understanding local institutions, nonetheless, takes time. Thus, in designing the project and gathering information on institutions, a researcher is often forced to choose between depth and breadth in their analysis. A case study of the institutional arrangements in just one community can often result in a rich fine grain analysis of the role of institutions in resource management (depth). If the goal, however, is to tease out broader patterns using quantitative methods (greater breadth), it can be hard to decide which institutions to focus on, and where to sacrifice depth (Poteete and Ostrom, 2008). In the study discussed here, we used a two-stage study that included mixed quantitative and qualitative methods to provide both breadth and depth to the analysis. That said, we still had to make decisions with respect to the granular level of analysis, as there were many land-use institutions that were not directly related to grazing behaviors that we were unable to include in the study. At the same time, we were also limited in breadth given time and resource constraints, the desire to have a deeper understanding of the role of existent communal governance conditions, and the resultant need to keep some contextual conditions constant.

Lastly, a challenge related to the above is the analytical tools available to assess the role of institutions in conjunction with the other contextual variables to explain the outcome of interest given field context limitations. Given the breadth of variables that can be considered under the IAD, and associated SES frameworks, a researcher can easily find themselves with too many explanatory variables, and too few cases. In our study, we initially used the IAD framework to identify a set of household, communal, biophysical, and institutional attributes that we intended to analyze using a set of regression analyses to test our hypotheses about factors influencing land-use behaviors. Given our fieldwork realities and limited number of cases, however, we could not use regression analysis to test all the models and variables we initially hoped to assess. That said, returning to the initial IAD framework variables and models we had constructed helped to identify how we could combine regression analysis with bivariate statistics and qualitative analysis to answer some of our institutional questions and understand the process of rule development, and ultimately, to consider the results within the broader socioeconomic, political, and environmental context.

SESs are complex and messy, and often filled with a maze of institutional arrangements. In working to tease out how institutions influence environmental management, we have found that the IAD provides a holistic structure to bring together various theories and data pieces to assess the links between conservation programs, collective decisions, local institutions, and behaviors in the commons.

Notes

1 This chapter draws from our published work on rule change (Hayes, Murtinho and Wolff, 2015) and behavioral change (Hayes, Murtinho and Wolff, 2017; Hayes et al., 2022). For those interested in applications of IAD to questions of participation and distribution of benefits, we refer readers to Hayes and Murtinho (2023) where we use the IAD to look at participation in communal decision-making processes and to Hayes and Murtinho (2018) that examines the distribution of PES costs and benefits across households.

2 Please see Hayes, Murtinho and Wolff (2017) and Hayes et al. (2022) for a full description of methods, as well as a discussion of the potential limitations.

3 This material comes from Hayes, Murtinho and Wolff (2015). Please see Hayes, Murtinho and Wolff (2015) for a more in-depth discussion of data gathering methods and the resultant findings.

4 The focus on access and use comes from the bundle of property rights defined by Schlager and Ostrom (1992). Specifically, we considered the creation of collective choice rules that define who can access the collective paramo lands, how they can be used, and how those rules are, in turn, monitored and enforced.

5 In our analyses, we were unable to run regressions as we were restricted by the sample size (there were only 47 participating communities in the highlands). Furthermore, while the 23 communities that had not joined PES allowed us to test for association between participation in PES and the development of land management institutions, given the potential for selection bias in the decision to join the PES program, we could not demonstrate causation.

6 For more information on the DID analysis and robustness tests that included the contextual variables, Hayes et al. (2022), Methods and Supplementary Table 3.

7 Given the limited number of communities in our sample, we were unable to use logit models to trace specifically how the creation of a new land-use rule in response to participating in PES was linked to behavioral change in the participant communities.

References

Agrawal, A. and Gibson, C.C. (1999) Enchantment and disenchantment: The role of community in natural resource conservation, *World Development*, 27(4): 629–649. https://doi.org/10.1016/S0305-750X(98)00161-2

Andersson, K., Benavides, J.P. and León, R. (2014) Institutional diversity and local forest governance, *Environmental Science & Policy*, 36: 61–72. https://doi.org/10.1016/j.envsci.2013.07.009

Andersson, K.P., Gibson, C.C. and Lehoucq, F. (2006) Municipal politics and forest governance: Comparative analysis of decentralization in Bolivia and Guatemala. *World Development*, 34(3): 576–595. https://doi.org/10.1016/j.worlddev.2005.08.009

Beedell, J. and Rehman, T. (2000) Using social-psychology models to understand farmers' conservation behaviour, *Journal of Rural Studies*, 16(1): 117–127. https://doi.org/10.1016/S0743-0167(99)00043-1

Bremer, L.L., Farley, K.A., Lopez-Carr, D. and Romero, J. (2014) Conservation and livelihood outcomes of payment for ecosystem services in the Ecuadorian Andes: What is the potential for 'win–win'? *Ecosystem Services*, 8: 148–165. https://doi.org/10.1016/j.ecoser.2014.03.007

Buytaert, W., Celleri, R., De Bievre, B., Cisneros, F., Wyseure, G., Deckers, J. and Hofstede, R. (2006) Human impact on the hydrology of the Andean paramos, *Earth-Science Reviews*, 79(1–2): 53–72. https://doi.org/10.1016/j.earscirev.2006.06.002

Cleaver, F. (2017) *Development through Bricolage: Rethinking Institutions for Natural Resource Management*, London: Routledge.

Cole, D.H. (2017) Laws, norms, and the Institutional Analysis and Development framework. *Journal of Institutional Economics*, 13(4): 829–847. https://doi.org/10.1017/S1744137417000030

Cole, D.H., Epstein, G. and McGinnis, M.D. (2019) Combining the IAD and SES frameworks, *International Journal of the Commons*, 13(1): 244–275. https://doi.org/10.18352/ijc.864

Cooke, B. and Kothari, U. (2001) *Participation: The New Tyranny?* London: Zed Books.

Crawford, S.E.S. and Ostrom, E. (1995) A grammar of institutions, *American Political Science Review*, 89(3): 582–600. https://doi.org/10.2307/2082975

Dietz, T., Ostrom, E. and Stern, P.C. (2003) The struggle to govern the commons, *Science* 302(5652): 1907–1912. https://doi.org/10.1126/science.1091015

Dougill, A.J., Stringer, L.C., Leventon, J., Riddell, M., Rueff, H., Spracklen, D.V. and Butt, E. (2012) Lessons from community-based payment for ecosystem service schemes: From forests to rangelands, *Philosophical Transactions of the Royal Society B-Biological Sciences*, 367(1606): 3178–3190. https://doi.org/10.1098/rstb.2011.0418

Ferraro, P.J. and Hanauer, M.M. (2014) Advances in measuring the environmental and social impacts of environmental programs, *Annual Review of Environment and Resources*, 39(1): 495–517. https://doi.org/10.1146/annurev-environ-101813-013230

Ferraro, P.J. and Pattanayak, S.K. (2006) Money for nothing? A call for empirical evaluation of biodiversity conservation investments, *PLoS Biology*, 4(4): e105. https://doi.org/10.1371/journal.pbio.0040105

Gibson, C.C., McKean, M.A. and Ostrom, E. (2000) Explaining deforestation: The role of local institutions, in Gibson, C.C., McKean, M.A. and Ostrom, E. (eds.) *People and Forests*, Cambridge: MIT Press.

Gibson, C., Williams, J. and Ostrom, E. (2003) Local enforcement and better forests, *World Development*, 33(2): 273–284. https://doi.org/10.1016/j.worlddev.2004.07.013

Grothmann, T. and Patt, A. (2005) Adaptive capacity and human cognition: The process of individual adaptation to climate change, *Global Environmental Change-Human and Policy Dimensions*, 15(3): 199–213. https://doi.org/10.1016/j.gloenvcha.2005.01.002

Hayes, T., Grillos, T., Bremer, L.L., Murtinho, F. and Shapiro, E. (2019) Collective PES: More than the sum of individual incentives, *Environmental Science & Policy*, 102: 1–8. https://doi.org/10.1016/j.envsci.2019.09.010

Hayes, T. and Murtinho, F. (2018) Communal governance, equity and payment for ecosystem services, *Land Use Policy*, 79: 123–136. https://doi.org/10.1016/j.landusepol.2018.08.001

Hayes, T. and Murtinho, F. (2023) Diagnosing participation and inclusion in collective decision-making in the commons: Lessons from Ecuador, *International Journal of the Commons*, 17(1): 37–53. https://doi.org/10.5334/ijc.1200

Hayes, T., Murtinho, F. and Wolff, H. (2015) An institutional analysis of payment for environmental services on collectively managed lands in Ecuador, *Ecological Economics*, 118: 81–89. https://doi.org/10.1016/j.ecolecon.2015.07.017

Hayes, T., Murtinho, F. and Wolff, H. (2017) The impact of payments for environmental services on communal lands: An analysis of the factors driving household land-use behavior in Ecuador, *World Development*, 93: 427–446. https://doi.org/10.1016/j.worlddev.2017.01.003

Hayes, T., Murtinho, F., Wolff, H., López-Sandoval, M.F. and Salazar, J. (2022) Effectiveness of payment for ecosystem services after loss and uncertainty of compensation, *Nature Sustainability*, 5(1): 81–88. https://doi.org/10.1038/s41893 021-00804-5

Jones, B.D. (1999) Bounded rationality, *Annual Review of Political Science*, 2(1): 297–321. https://doi.org/10.1146/annurev.polisci.2.1.297

Kerr, J.M., Vardhan, M. and Jindal, R. (2014) Incentives, conditionality and collective action in payment for environmental services, *International Journal of the Commons*, 8(2): 595–616. https://doi.org/10.18352/ijc.438

Koontz, T.M. (2001) Money talks? But to whom? Financial versus nonmonetary motivations in land use decisions, *Society & Natural Resources*, 14(1): 51–65. https://doi.org/10.1080/08941920117246

Kosoy, N., Corbera, E. and Brown, K. (2008) Participation in payments for ecosystem services: Case studies from the Lacandon rainforest, Mexico, *Geoforum*, 39(6): 2073–2083. https://doi.org/10.1016/j.geoforum.2008.08.007

Larson, A.M. and Soto, F. (2008) Decentralization of natural resource governance regimes, *Annual Review of Environment and Resources*, 33(1): 213–239. https://doi.org/10.1146/annurev.environ.33.020607.095522

Ministerio del Ambiente (2009) *Acuerdo Ministerial Numero 115*, Ministerio del Ambiente, Quito, Ecuador.

Meinzen-Dick, R. (2007) Beyond panaceas in water institutions, *Proceedings of the National Academy of Sciences of the United States of America*, 104(39): 15200–15205. https://doi.org/10.1073/pnas.0702296104

Muradian, R. (2013) Payments for ecosystem services as incentives for collective action. *Society & Natural Resources*, 26(10): 1155–1169. https://doi.org/10.1080/08941920.2013.820816

Murtinho, F. and Hayes, T. (2017) Communal participation in payment for environmental services (PES): Unpacking the collective decision to enroll, *Environmental Management*, 59(6): 939–955. https://doi.org/10.1007/s00267-017-0838-z

Ostrom, E. (1990) *Governing the Commons: The Evolution of Institutions for Collective Action*, Cambridge: Cambridge University Press.

Ostrom, E. (2005) *Understanding Institutional Diversity*, Princeton, NJ: Princeton University Press.

Ostrom, E. (2009) A general framework for analyzing sustainability of social-ecological systems, *Science*, 325(24): 419–422. https://doi.org/10.1126/science.1172133

Ostrom, E. (2011) Background on the institutional analysis and development framework, *Policy Studies Journal*, 39(1): 7–27. https://doi.org/10.1111/j.1541-0072.2010.00394.x

Ostrom, E., Agrawal, A., Blomquist, W., Schlager, E. and Tang, S.Y. (1989) *Common-pool Resources (CPR) Coding Manual*, Bloomington, IL: Workshop on Political Theory and Policy Analysis.

Ostrom, E. and Cox, M. (2010) Moving beyond panaceas: A multi-tiered diagnostic approach for social-ecological analysis, *Environmental Conservation*, 37(4): 451–463. https://doi.org/10.1017/S0376892910000834

Ostrom, E. and Nagendra, H. (2006) Insights on linking forests, trees, and people from the air, on the ground, and in the laboratory, *Proceedings of the National Academy of Sciences*, 103(51): 19224–19231. https://doi.org/10.1073/pnas.0607962103

Pascual, U., Phelps, J., Garmendia, E., Brown, K., Corbera, E., Martin, A., Gomez-Baggethun, E. and Muradian, R. (2014) Social equity matters in payments for ecosystem services, *Bioscience*, 64(11): 1027–1036. https://doi.org/10.1093/biosci/biu146

Poteete, A. and Ostrom, E. (2004) Heterogeneity, group size and collective action: The role of institutions in forest management, *Development and Change*, 35(3): 435–461. https://doi.org/10.1111/j.1467-7660.2004.00360.x

Poteete, A.R. and Ostrom, E. (2008) Fifteen years of empirical research on collective action in natural resource management: Struggling to build large-N databases based on qualitative research, *World Development*, 36(1): 176–195. https://doi.org/10.1016/j.worlddev.2007.02.012

Pynegar, E.L., Gibbons, J.M., Asquith, N.M. and Jones, J.P.G. (2021) What role should randomized control trials play in providing the evidence base for conservation? *Oryx*, 55(2): 235–244. https://doi.org/10.1017/S0030605319000188

Raphael, K. (1987) Recall bias: A proposal for assessment and control, *International Journal of Epidemiology*, 16(2): 167–170. https://doi.org/10.1093/ije/16.2.167

Ribot, J.C. (2003) Democratic decentralization of natural resources, in Van De Walle, N., Ball, N. and Ramachandran, V. (eds.) *Beyond Structural Adjustment*, New York: Palgrave Macmillan, pp. 159–182. https://doi.org/10.1057/9781403981288_6

Schaeffer, N.C. and Presser, S. (2003) The science of asking questions, *Annual Review of Sociology*, 29: 65–88. https://doi.org/10.1146/annurev.soc.29.110702.110112

Schlager, E. and Ostrom, E. (1992) Property-rights regimes and natural-resources—A conceptual analysis, *Land Economics*, 68(3): 249–262. https://doi.org/10.2307/3146375

Shadish, W.R., Cook, T.D. and Campbell, D.T. (2002) *Experimental and Quasi-Experimental Designs for Generalized Causal Inference*, Boston, MA: Houghton Mifflin.

Thornberry, T.P. and Krohn, M.D. (2000) The self-report method for measuring delinquency and crime, *Criminal Justice*, 4(1): 33–83.

Tourangeau, R. and Yan, T. (2007) Sensitive questions in surveys, *Psychological Bulletin*, 133(5): 859–883. https://doi.org/10.1037/0033-2909.133.5.859

Vatn, A. (2010) An institutional analysis of payments for environmental services, *Ecological Economics*, 69(6): 1245–1252. https://doi.org/10.1016/j.ecolecon.2009.11.018

Vignola, R., Koellner, T., Scholz, R.W. and McDaniels, T.L. (2010) Decision-making by farmers regarding ecosystem services: Factors affecting soil conservation efforts in Costa Rica, *Land Use Policy*, 27(4): 1132–1142. https://doi.org/10.1016/j.landusepol.2010.03.003

3 Applying the institutional grammar in the study of natural resource governance

Saba Siddiki and Edella Schlager

Introduction

Policy and Institutional analysts have had a long-standing interest in evaluating the designs and performance of institutional arrangements engaged by communities in natural resource governance. Institutional arrangements allow people and organizations to resolve collective action problems and coordinate and cooperate in using, managing, and governing natural resources. One of the most prominent approaches used to guide such evaluations is the Institutional Analysis and Development (IAD) framework (Ostrom, 2005). In addition to offering theoretical and conceptual guidance to support understanding of how institutions guide and constrain individual and collective action, the IAD framework also offers several analytical approaches that enable systematic analysis (Schlager and Villamayor Tomas, 2023).

This chapter introduces one such approach, called the Institutional Grammar (IG) (Crawford and Ostrom, 1995; Frantz and Siddiki, 2022). The IG allows policy analysts to examine the syntactic components constituting prescriptive statements that comprise laws, policies, and regulations. Such granular analyses open a range of natural resource governance topics for study, ranging from forms of decision-making, the exercise of discretion, and the legitimacy of governing arrangements, to the design of monitoring and enforcement mechanisms, and the means of constituting and revising natural resource governing arrangements.

As part of this introduction, this chapter (i) provides a general overview of the IG; (ii) offers guidance on how it can be applied along with a set of qualitative, quantitative, and computational analysis techniques; and (iii) presents a brief review of select applications of the IG to the study of natural resource governance that exemplify the use and utility of employing the IG along with these various techniques. First, we begin by situating the IG within the larger context of the IAD framework.

The institutional analysis and development framework and related conceptual tools for institutional analysis

The IAD framework, developed by Elinor Ostrom and colleagues (Kiser and Ostrom, 1982; Ostrom, 2005), has played a prominent role in the study

DOI: 10.4324/9781032624679-3

of natural resource governance over the past four decades (Ostrom, 2005; Ostrom et al., 1994; Poteete et al., 2010; Schlager and Villamayor Tomas, 2023). The IAD framework is problem oriented. It provides policy analysts and policy practitioners with conceptual guidance for studying situations in which people confront shared problems and attempt to address and resolve them by developing an understanding of the issues involved and devising institutional arrangements well matched to the social and ecological context. The framework is both diagnostic and prescriptive.

The problem orientation of the framework makes it especially well-suited for examining natural resources dilemmas as well as the institutional arrangements intended to mitigate them. Institutional analysts can use the framework to explore a wide range of topics. What types of collective choice and constitutional choice venues are in place that people may access and participate in to create institutional arrangements or revise existing policies and regulations? Who is allowed to participate in such venues and what types of authority do they have? What general institutional designs or features are more effective in supporting coordination, cooperation among natural resource users and decision makers, as well as how do institutional designs encourage compliance? In sum, the IAD framework, with its focus on institutional arrangements in context, allows scholars to study how institutional arrangements structure people's choices and activities, and in turn how people's choices structure institutional arrangements.

Within the IAD framework, institutional arrangements are understood as configurations of institutional statements (Ostrom, 2005). Institutional statements describe expected actions for actors within the presence or absence of constraints, or parameterized features of an institutional system (Frantz and Siddiki, 2022). Institutional arrangements can only be understood within the context in which they operate, and that context includes the biophysical, the community, and the institutional arrangement settings within which they are nested. People use rules as tools to address collective action dilemmas that emerge in the provision and governance of public goods and common pool resources (Ostrom, 2005). The resulting governing arrangements, or configurations of institutional statements, affect the coordination, competition, conflict, and cooperation of actors in interdependent situations (Ostrom and Basurto, 2011). How effective the governing arrangements are at addressing collective action dilemmas and permitting the sustainable use of common pool resources depends on how well they fit the setting in which they are used and reflect shared conceptions of equity, especially in the distribution of benefits and costs (Ostrom 2005; Baggio, et al. 2016). As Ostrom (2005: 3) notes, 'if the individuals who are crafting and modifying rules do not understand how particular combinations of rules affect actions and outcomes in a particular ecological and cultural environment, rule changes may produce unexpected and, at times, disastrous outcomes'.

The initial IAD framework provides institutional and policy analysts with multiple analytical tools for systematically organizing and studying configurations of institutional arrangements and institutional statements in context (Ostrom, 1999; Ostrom et al., 1989, 1996). The concept of the 'action situation' allows analysts to identify the focal units of study. Action situations are 'a social space where participants with diverse preferences interact' (Ostrom, 2005: 14). Action situations consist of actors, in positions, who have the authority to act, or make choices, sometimes individually and sometimes with others, based on information as well as the consequences of such choices. The action situation takes place in a context structured by a material and/or biophysical milieu, the community setting, and rules in use and form. The material and biophysical setting is further organized by the goods typology introduced by Ostrom and Ostrom (1977). The goods typology motivates the social dilemmas that actors in action situations experience and attempt to resolve. The typology consists of common pool resources and public goods, whereby it is costly to exclude people from either type, but common pool resources are subtractable whereas public goods are not, at least over a range of uses. The typology also includes private goods and toll goods, whereby exclusion is feasible, and private goods are subtractable whereas toll goods are not. The community setting includes culture, world views, and social norms. The community setting shapes how people view one another and interact with each other. Finally, the rules include those that are in use, that is, the rules that individuals are using and following in the action situation. They may or may not be identical with the rules in form, which are rules that have been adopted by a recognized, authoritative venue, such as a council, board, or legislature (Ostrom, 2005). With the levels of action and the evaluative criteria, the IAD framework provides scholars with a rich set of concepts and analytical tools to study governance, as illustrated by the common pool resource research program (Ostrom, 1990) and Elinor Ostrom's award of a Nobel Prize in Economics in 2009.

Over time, additional analytic tools were developed and incorporated within the IAD framework, supporting the emergence of new lines of research on institutional arrangements (Schlager and Villamayor Tomas, 2023). The property rights typology (Schlager and Ostrom, 1992), networks of adjacent action situations (McGinnis, 2011), and the SES framework (Ostrom, 2009) provided approaches that allowed scholars to systematically examine nested governance as well as the governance of regional and global common pool resources that cross jurisdictional boundaries (Leslie et al., 2015; Schlager and Heikkila, 2011; McCord et al., 2016).

Of note, and the focus of this chapter, is the Institutional Grammar (IG). The IG orients the study of institutional arrangements on the analysis of institutional statements and offers a structured approach for evaluating the structure (i.e., syntax) and meaning (i.e., semantics) of institutional statements. The IG also offers analytical guidance on how to examine actors' valuation of compliance with statements. The Grammar, combined with the rule typology, allows analysts to methodically examine the institutional structure of action situations. Ostrom (2005) proposed using the 'action' component of the Grammar to identify rule types. Rule types structure the components of action situations (Ostrom, 2005). McGinnis (2011) proposed using the components of action situations as one means of identifying how action situations are linked. Thus, the Grammar, the rule typology, and action situations together provide policy scholars and analysts with the means of addressing a wide range of natural resource governance questions.

The following section provides an in-depth description of the IG, how to use it to code institutional statements and how to use Grammar-coded data sets to empirically test natural resource governance questions.

The institutional grammar

The IG, as an approach embedded within the IAD framework to enable reliable and rigorous assessments of the design of institutional arrangements, can be applied to study institutions-in-form or institutions-in-use that govern social settings. Of the different institutional analysis approaches affiliated with the IAD framework, the IG provides the most granular perspective and associated analytical guidance on the study of institutional arrangements. As noted in the previous section, the central unit of analysis in the IG is an institutional statement, which can, and typically does, configure with other institutional statements to form institutional arrangements. The IG furthermore identifies common features (i.e., syntactic components each bearing institutionally relevant meaning) along which institutional statements can be parsed and evaluated.

A brief introduction

The IG was originally proposed by Crawford and Ostrom (1995) to aid in the definition and analysis of institutions. Crawford and Ostrom proposed clarifying what institutions are, based on the kinds and extent of behavioral prescription they convey. The task of defining and differentiating among different types of institutions is made easier because institutions convey this information through common structural forms (i.e., constellations of syntactic components), which can be conceptualized with an 'IG'. Crawford and Ostrom noted that institutional statements convey at least three types of information: an actor (individual or corporate) that carries out or is expected to/to not carry out an action described in the statement, the goal or action of the statement assigned to the statement actor, and conditions or constraints on actions. Institutional statements can further express various degrees of pro/prescription of actions and incentives for statement (non)compliance.

Crawford and Ostrom represented these kinds of information, all of which bear relevance for understanding how institutions govern behavior, with unique syntactic components which combine to form the IG. The original IG syntax, as proposed by Crawford and Ostrom, has been revised over the last 30 years, to resolve ontological inconsistencies, better capture the diversity of institutional statements encountered, and support the analysis of institutional statements using a diverse array of qualitative, quantitative, and computational methods (Frantz and Siddiki, 2022).

The most recent iteration of the IG, referred to as the IG 2.0, differentiates between 'regulative' and 'constitutive' statements (Frantz and Siddiki, 2022). Regulative statements describe actors' duties and discretion linked to specific actions within certain contextual parameters. Constitutive statements parameterize an institutional setting, and in the process, introduce, modify, or otherwise constitute features of an institutional system and action situations embedded therein, including aspects such as actor positions and roles, associated affordances, processes, environmental characteristics, objects, and artifacts in as much as they are institutionally relevant. Taken together, regulative and constitutive statements convey, through their design, behaviors within institutionally governed systems as well as attributes of those systems in which, through which, or on which behaviors occur.

Regulative statements are composed of some or all of the following components with the corresponding syntactic labels and symbols (indicated in parentheses): (i) a responsible actor, referred to as an Attribute (A), (ii) action regulated by the statement, referred to as an Aim (I); (iii) statement context, referred to as Context (C); (iv) a receiver of action, referred to as an Object (B); (v) a prescriptive operator that describes how strongly an action is compelled or restrained, referred to as a Deontic (D); and (vi) a consequence of violating the regulated action, referred to as an Or else (O). The Deontic is used to distinguish between prescriptive and non-prescriptive statements and is typically represented by 'may' (permitted), 'must' (required), or 'must not' (forbidden) (Ostrom, 2005: 140). The Context, in regulative and constitutive statements (the latter are described next), can be further delineated into Activation Conditions (Cac) and Execution Constraints (Cex). Activation conditions are kinds of Context clauses that instantiate settings in which the focal action of the statement applies. Execution constraints qualify the action of the statement. Additionally, Attributes and Objects (in regulative statements) can be further delineated to account for descriptors of each.

An example of a regulative statement is: 'The City shall develop and implement a Watershed Forest Conservation Easement Program within 12 months from the effective date of the permit'.

In this statement, the 'City' is the Attribute, 'shall' is the Deontic, 'develop' and 'implement' are Aims, 'Watershed Forest Conservation Easement Program' is the Object, and 'within 12 months from the effective date of the permit' is the execution constraint. The statement does not contain an Or else. Inline coding of institutional statements organizes statement information within parentheses preceded by syntactic symbols. The regulative statement example would be coded as follows [see Frantz and Siddiki (2022) for complete description of IG script and inline coding]:

The A(City) D(shall) I((develop [AND] implement)) a O(Watershed Forest Conservation Easement Program) Cex(within 12 months from the effective date of the permit).

Constitutive statements are composed of some or all of the following components with the corresponding syntactic labels and symbols (indicated in parentheses): (i) the entity that is being constituted or directly modified within a statement, referred to as a Constituted Entity (E); (ii) a parameterizing activity or function that introduces or otherwise characterizes the Constituted Entity in relation to the institutional setting and potential Constituting Properties, called the Constitutive Function (F); (iii) the statement context, referred to as Context (C); (iv) properties that serve as input to the constitutive function, called Constituting Properties (P); (v) a modal operator that defines to which extent the constitutive function of an institutional statement is required (necessary) or merely possible (optional), referred to as a Modal (M); and (vi) a consequence linked to the non-fulfillment of the function referenced in the constitutive function, referred to as an Or else (O). Entities and Constituting Properties can be further delineated to account for descriptors of each.

An example of a constitutive statement is: 'Land trusts shall be bound to comply with the provisions of this permit by contract'.

In this statement, 'Land trusts' are the Constituted Entity, 'shall' is the Modal, 'be' is the Constitutive Function, 'bound to comply with the provisions of this permit' is the Constituting Property, and 'by contract' is the execution constraint. This statement does not contain an Or else. The inline coding of the example constitutive statement is as follows:

E(Land trusts) M(shall) F(be) P(bound to comply with the provisions of this permit) Cex(by contract).

Conceptually, the decomposition of institutional statements by syntactic components affords the ability to understand common structure and meaning of institutional statements. From a methodological standpoint, this decomposition offers a clear unit of analysis to enable rigorous and reliable assessments of institutional design. Namely, the IG orients the study of institutional configurations on the analysis of institutional statements and their constituent parts. In this way, the IG enables assessment of institutional statement level and/or syntactic component level.

Statement level analyses focus on patterns in qualities or types of institutional statements. For example, they may focus on deciphering the relative share of constitutive to regulative statements within an institutional arrangement toward understanding of whether it is primarily focused on defining the parameters of an institutional setting, or the behavior observed therein. Additionally, statement level analyses may also involve the classification of statements in accordance with other institutional typologies affiliated with the IAD framework. For example, scholars may use the IG to engage in a second-order classification of institutional statements according to the IAD's rule typology. Crawford and Ostrom (2005) reference this application describing how syntactic information can be leveraged in the identification of rule type. They specifically note that information communicated in the Aim of an institutional statement can be used to signal the functional properties of institutional statements (i.e., the type of 'rule' that an institutional statement would classify as). Applying the IG in conjunction with the rule typology in the study of institutional arrangements lends measurement in clarity inasmuch as the IG provides a clear unit of analysis (i.e., institutional statements) that forms the basis of the rule type classification. This, in turn, fosters reliability in the rule type coding.

Syntactic level analyses commonly entail examinations of patterns in information corresponding to one or more syntactic components across institutional statements. Such assessments have been used in existing IG applications to understand the array of actors, actions, and objects present across some set of institutional statements. They have also been used to detect and derive understanding from conceptually relevant patterns in how syntactic values link together. For example, in seeking to better understand, how discretion is variably afforded to different types of actors present within an institutional arrangement, one can analyze systematic differences in the Deontics used in relation to actions assigned to specific kinds of institutional actors. As another example, toward better understanding of institutional stringency, the analyst can examine how often, and with respect to which activities, are Deontics paired within specific or else within and across institutional statements.

Syntactic-level analyses have also been used to understand 'modal' syntactic values, such as the most common actors, actions, and objects conveyed within a set of statements. These kinds of assessment support understanding of the foci of institutional arrangements, for example. Scholars have also increasingly used syntactic information in conjunction with network analysis methods to understand how information conveyed within and across institutional statements connects. This approach, in essence, supports the construction of 'institutional networks', thereby conveying the importance of analyzing institutional statements in configural terms rather than in isolation. Each of these types of applications is discussed in more detail later in this chapter. Finally, as implicitly conveyed here, and explicitly by Frantz and Siddiki (2022), applications of the IG do not require either the coding or analysis of all syntactic components of regulative or constitutive statements. The analyst may choose to focus on select components.

Scholars employing the IG often pair statement and syntactic analyses in their studies, recognizing not only distinctive but also complementary insights they enable. However, fundamentally, the type of analysis one engages is dependent on his/her research question.

Levels of expressiveness

The recently developed IG 2.0 affords additional analytical possibilities stemming from the application of the IG to study institutional arrangements. The IG 2.0 facilitates coding of institutional statements at different 'levels of expressiveness', referred to as IG Core, IG Extended, and IG Logico. According to Frantz and Siddiki, institutional statements can be coded at different levels of expressiveness to draw out varying degrees of structural (i.e., syntactic) and semantic (i.e., meaning) detail and/or complexity. Additionally, institutional statements can be coded in a way that allows for explicit connection of syntactic information to concepts or theory of interest. Coding for richer characterization of structural and semantic detail and complexity of institutional statements also supports the use of computational methods in the collection and modeling of institutional information. Complete descriptions of the different levels of expressiveness, and associated coding, can be found in Siddiki and Frantz (2021) and Frantz and Siddiki (2022). A concise overview of each level and associated analytical opportunities is offered here.

Coding of institutional statements at the IG Core level involves the most basic structural parsing of statements. Statements are decomposed along regulative/constitutive components, while also capturing how multiple values associated with a single syntactic component are logically linked (i.e., via AND, OR, XOR operators) within institutional statements. Reference to the example regulative statement provided earlier in this section offers an illustration of the latter. In the example statement, there are two referenced activities, 'develop' and 'implement', that are coded as statement Aims. These two actions are explicitly combined with the logical operator 'and', substantively conveying that the Attribute of the statement (i.e., the City) must perform both activities.

IG Extended coding of institutional information supports a more fine-grained depiction of institutional statement information. Coding at this level enables comprehensive representation of the information explicitly and implicitly conveyed in institutional statements as well as supports identification of second-order properties where they exist.

To illustrate IG Extended coding, consider the revised example of the regulative statement referenced throughout this section: 'Once the State issues guidance, the City shall develop and implement a revised, local Watershed Forest Conservation Easement Program within 12 months from the effective date of the permit'. The original regulative statement example has been augmented with an activation condition – 'Once the state issues guidance' – and two qualifications of the Watershed Forest Conservation Easement Program: 'revised' and 'local'. A basic structural parsing of this statement, as would be undertaken if coding it at the IG Core level, would simply beg annotation of the added Context clause as an activation condition and the added qualifiers of the Program as Object properties. An IG Extended coding, given that it prompts explicit coding of implicit and explicit information conveyed in institutional statements, would go one step further to also identify the implied action conveyed in the activation condition: necessary approval by the state agency. This implied action would be conveyed as its own institutional statement but coded and nested within the original statement. An IG Extended coding would further prompt identification of properties of multiple order (i.e., first- and second-order properties). Both would be captured in the revised statement coded as follows:

{Cac(Once the A(State) I(issues) O(guidance)}, the A(City) D(shall) I((develop [AND] implement)) a O,p1(revised) O,p2(local) O(Watershed Forest Conservation Easement Program) Cex(within 12 months from the effective date of the permit.

Fundamentally, IG Extended coding of institutional statements is intended to enhance the ability of computers to effectively parse statements along IG syntactic categories as well as the usability of IG coded data in computational modeling (e.g., agent-based modeling). Accurate computer-assisted coding of statements is enabled by parsimonious delineation of syntactic components. Agent-based models, if the design of which is premised on information communicated within institutional statements, require complete representations of actors and actions explicitly and implicitly communicated therein. Hence, construction of such models is aided by IG Extended coding of institutional statements that provide a structured approach for characterizing the variably explicit actions relevant within a particular institutional setting.

Finally, IG Logico coding supports semantic labeling and analysis of institutional statements, explicit referencing of related institutional statements, and ability to perform logical transformations on institutional statements. Here semantic labeling refers to annotating institutional information corresponding to different syntactic categories with conceptually relevant labels. The objective of this annotation is to aid in the connection and interpretation of institutional information in conceptually or theoretically meaningful ways.

For illustration, consider again the regulative statement example: 'The City shall develop and implement a Watershed Forest Conservation Easement Program within 12 months from the effective date of the permit'. In addition to parsing this statement syntactically, the analyst may wish to explicitly annotate the role of the Attribute in the institutional setting. For example, it is clear in this case that the City is an implementing agent with respect to the referenced program. If the analyst wanted to capture this role in the institutional statement coding, to assist in the conceptual or theoretical organization of institutional statement information and/or the downstream processing thereof, the statement could be annotated as follows alongside the typical syntactic parsing of the statement:

The A [role=implementing agent](City) D(shall) I((develop [AND] implement)) a O(Watershed Forest Conservation Easement Program) Cex(within 12 months from the effective date of the permit).

Semantic labeling can also be used to capture institutional statement linkages, inasmuch as labels can be used to make reference to other institutional statements to which the statement being coded is linked. For example, when coding institutional arrangements found in policy documents, it is not uncommon to find institutional statements that reference information contained in other parts of the policy, or even in different related policies. Semantic labeling can be used to catalog information about other policy sections or policies, the understanding of which is important for contextualizing or interpreting the information being conveyed in the institutional statement under consideration.

The final faculty afforded by IG Logico is logical transformation of institutional statement information. The IG 2.0 offers specific guidance on how the institutional analyst can reconstruct institutional statements to reflect their logical alternatives in case this kind of reconstruction is valuable for one's analytical objectives.

Applying the institutional grammar

The preceding section offers a conceptual overview of the IG. In complement, this section offers guidance on conventional ways that the IG is applied. As with any institutional study, a study involving an application of the IG begins with the identification of a focal collective action dilemma and related institutional context. Collective action dilemmas are undesirable social outcomes arising from patterns of interaction among actors. Collective action dilemmas emerge or result from contexts, including institutional contexts. An initial task in an IG study is the identification of institutions (in-form and/or in-use) that are presumably most proximately linked to the dilemma of interest and to which the IG can be applied.

Another complementary initial task for the institutional analyst is the selection of theory and research questions to guide inquiry about the institutional contexts of the collective action dilemma of interest. Identifying research questions at the outset of an IG study is particularly important as it will guide how one engages in the coding of institutions, including at what level of expressiveness coding will be conducted as well as what syntactic components will be coded for. Later in the research process, guiding theory and research questions will inform how data are analyzed; for example, whether institutional information will be analyzed at the statement or syntactic component level and what specific patterns in statements and/or components the analyst will focus on detecting and interpreting.

Generic resources (e.g., codebooks, coding tutorials) exist that offer comprehensive and detailed guidance on the IG coding process (these resources are available at: https://institutionalgrammar.org/). These resources offer general guidance on how to apply the IG. Alongside these resources, individual researchers may wish to develop project specific codebooks that offer coding guidance relating to qualities of institutions being analyzed in their specific projects. For example, institutions-in-form such as public policies may have structural or stylistic features that have implications for how institutional information is coded and relating to which researchers may wish to develop project-specific coding heuristics. Similarly, oral communications of institutional statements may require coding rules or heuristics to aid both in the identification of institutional statements embedded in oral transcripts as well as in the coding of these institutional statements. Project-specific codebooks can be particularly advantageous when research projects involve multiple coders. In this case, it is helpful if these codebooks are co-developed by all coders who will be involved in the IG coding process in an iterative fashion, that is, based on multiple iterations of coding of an institutional sample and related application/refinement of coding rules. Ultimately, drawing on a project specific codebook along with general coding guidance will ensure intra- and inter-coder reliability.

Returning to more general guidance on IG coding, offered here are conventional practices in IG coding that are more fully detailed in the references provided earlier in this section. Generally, once institutions of interest have been identified and collected, those institutions (in entirety or selective sampling) are reviewed to gain a high-level understanding of institutional actors, actors, and institutional statement relationships that can be leveraged in the coding process. Next, the analyst selects the coding methodology (e.g., manual or computational) and related tooling or software. Once these selections have been made, the institutional analyst can officially embark on the coding process in accordance with general and project-specific coding guidelines. Some pre-processing of institutions may be useful to remove content or stylistic features (e.g., punctuation) that are irrelevant for understanding institutional patterns of interest or that might compromise the validity or reliability of the coding. Following this pre-processing, the institutional analyst parses the institutions, either comprehensively or in multiple passes. With the former, institutions would be fully decomposed into regulative and constitutive statements and related syntactic components in a first coding pass. With the latter, the analyst may first choose to parse institutions into regulative and constitutive statements in a first pass, and then by syntactic components in a second pass (though some basic, even if implicit, parsing of statements is required to accurately categorize them as regulative or constitutive).

Applying the IG in the study of natural resource governance

Over the past 15 years, the IG has been applied to study governance in a wide variety of topical domains, including food systems, health, local decision-making, among others [see Siddiki et al. (2022) and Frantz and Siddiki (2022) for recent reviews of existing research]. A significant share of extant IG applications involves the study of natural resource governance. These studies are often informed by the IAD framework and related theory (e.g., Common Pool Resource Theory), and also engage other theories from political science, public administration, policy studies, economics, and cognate fields. Furthermore, these studies use the IG in concert with a variety of methodological techniques as discussed later in this chapter. Despite drawing on different theories and methods, there is overlap in the kinds of research questions that scholars are responding to in their IG research. Among these are the following:

1. Which actors are assigned responsibility for managing natural resources?
2. What incentives are conveyed through institutions for cooperation/coordination in natural resource governance?
3. What incentives are conveyed through institutions to foster institutional compliance?
4. How do institutions prescribe/describe collective decision making in natural resource governance?
5. How do institutions-in-form reflect institutions-in-use and vice versa?
6. How do institutions governing different types of public goods dilemmas differ in design?

Conceptual and empirical IG research is similarly aimed at understanding actors involved in natural resource governance, how these actors cooperate, the variable choices and discretion afforded to them, the behavioral incentives they face, and characteristics of the systems in which they act and interact. Among the questions that institutional scholars could pursue in future research through application of the IG are the following:

1. What natural resource governance goals are conveyed in institutions?
2. How does power embed in institutional design?
3. How do institutions define the boundaries of natural resource governance systems?
4. How are natural resource governance systems constituted?
5. How do institutions create institutional facts and realities that shape both the structure of natural resource governance systems and the behavior observed therein?

In the following section, we offer an elaborated discussion of how IG coded data can be analyzed drawing on a range of qualitative, quantitative, and computational techniques referencing select applications.

Analyzing institutional grammar data

With growing use of the IG, there has been an increase in the diversity of methodological approaches with which scholars are analyzing IG data. Most commonly, however, scholars are engaging in (i) descriptive (qualitative and quantitative) analysis of IG data with the objective of identifying general conceptually relevant trends observed therein; (ii) statistical modeling of IG data toward diagnostic evaluation of antecedents or outcomes of institutional design; (iii) network modeling of IG data primarily for detecting, visualizing, and measuring qualities of institutional configurations; and (iv) computational (i.e., agent-based modeling) of IG data with the intent of studying emergent outcomes associated with IG-based institutional scenarios or post hoc interpretation of emergent behaviors. In addition to being the most common approaches for analyzing IG data generally, these are also the approaches most observed in IG studies of natural resource governance.

In the remainder of this section, we offer more detail about what descriptive, statistical, network, and computational analysis of IG data entails. As part of the forthcoming discussion, we also reference specific natural resource governance IG studies that employ the referenced analytical approaches.

Descriptive analysis of IG data

One of the most common uses of Grammar coded data is to identify types of institutional designs or categories of institutional arrangements and to use the categories to compare cases or to compare a case against an ideal type (Pieper et al., 2023). One of the most common uses of the IG is to distinguish among rules, norms, and strategies. Scholars identify rules, norms, and strategies by the presence or absence of components of the Grammar as defined by Ostrom and Crawford (1995): (1) rule: an institutional statement that contains all components of the Grammar – Attributes, Deontic, Aim, Object, Condition, and Or else (see prior section); (2) norm: an institutional statement composed of the Attributes, Deontic, Aim, Object, and Condition; (3) strategy: an institutional statement that contains three of the components – Attributes, Aim, and Condition.

For instance, Novo and Garrido (2014) examine the Nicaragua Water Law, adopted in 2007 and the barriers encountered in its implementation. In addition to relying on 40 in-depth interviews of mostly public sector actors involved in water management, the authors also Grammar-coded the law and used the coded information to identify rules, norms, and strategies. They presented the data as raw counts of the different types of statements by Water Law titles, water administration, and water uses (see Tables 5–7, pp. 1025–1027). This allowed the authors to compare the roles of administrative bodies by the norms and strategies assigned them, as well as to examine the types of water uses that received the most attention under the law. Using rules, norms and strategies as categories to compare the designs of different types of laws is common in studies of natural resource governance (see Watkins and Westphal (2015) on governance of urban forest reserves; and Clement et al. (2015) on land conservation for biodiversity, for examples).

Grammar-coded data has also been used to construct variables representing institutional arrangements that are theoretically meaningful. The variables are used to empirically test hypotheses regarding institutional designs. For instance, Hanlon, Olivier and Schlager (2019) identified three types of arrangements used by actors to hold one another accountable for following the rules they adopted for protecting water quality and enhancing environmentally sensitive economic development through the provision of public goods in the Delaware and Catskill watersheds from which New York City sources its drinking water. For each public good, the three types of arrangements, monitoring, reviewing, and sanctioning, were identified (Hanlon, Olivier and Schlager, 2019: 982). Each type of arrangement consisted of 1-N consecutive institutional statements. Most such arrangements are not reflected in a single type of institutional statement. For instance, a monitoring mechanism typically does not consist of a single information rule; rather, it may include a choice rule, authorizing an actor to gather data, or a position rule creating the role of monitor. Likewise with reviewing arrangements.

Each statement was classified by the rule typology and Grammar-coded. The coding allowed the authors to identify the number of decision points using the aggregation rules. In addition, using the Attributes and Object components, which actors were monitoring and being monitored, and which actors were reviewing and being reviewed could be identified. Sanctions and their severity were qualitatively determined by the content of the statements. Hanlon, Olivier and Schlager (2019) compared the presence and design of the accountability arrangements between water quality and economic development public goods using frequencies and *t*-tests to examine differences in means.

Statistical analysis with IG data

Statistical analysis of IG data involves the descriptive summary of IG data represented in numerical terms, or the use of numerically represented IG data in statistical models (Chen et al., 2022; Schlager et al., 2021). The former typically entails assessment of the frequency with which certain qualitative values corresponding with different syntactic components appear across a set of institutional statements. For example, an analyst may be interested in identifying how many times a particular actor appears as the Attributes of statements comprising an institutional arrangement. It may also entail descriptive summaries of such values which have already been translated into numerical values. For example, in an analysis of institutional stringency, the analyst could numerically capture whether a statement contains a stringent deontic (e.g., 'must' or 'must not'), representing each instance of such with a value of '1'. The analyst could then use these values to descriptively represent the aggregate level of stringency across a set of institutional statements.

The translation of IG coded data to numerical values to aid in the quantitative description of statement information also aids in the statistical modeling of IG data. With statistical modeling of IG data, the institutional analyst can examine factors contributing to observed statement values. The analyst can additionally examine outcomes associated with institutional arrangements bearing particular designs. Key considerations in the statistical analysis of IG data, include: (i) ensuring that numerical representations of the raw, qualitative values generated through IG coding are substantively valid and (ii) ensuring, particularly in the case of statistical modeling, that the analyst has enough observations (e.g., statements, syntactic values) to support robust statistical analysis.

Recent IG studies of natural resource governance highlight the value of employing statistical methods in the analysis of IG data. For example, Schlager et al. (2021) use an IG coding of institutional statements found in water governance documents as part of a logistic regression modeling exercise that aims to examine whether institutional arrangements designed to address different collective action dilemmas vary. As part of their analysis, they use the IG to identify specific features of institutional statements along with how those associated with different collective action dilemmas may differ. In another recent study, Chen et al. (2022) used an IG inspired coding of institutional information captured in oil and gas regulation as part of a Cox proportional hazard modeling exercise that aims to examine whether more stringent regulation takes longer to get adopted. They characterize stringency at the policy level, rather than institutional statement level, based on the automated extraction of Deontics conveyed within policy text.

Agent-based modeling with IG data

The earliest applications of the IG, following its introduction by Crawford and Ostrom, involved its integrated use with agent-based modeling (ABM) (Smajgl, Izquierdo and Huigen, 2008). In existing research, the IG is used toward the upfront parameterization of models (i.e., toward the specification of agents, agent actions, and system features being reconstructed in the modeled environment) or in the post hoc description of behaviors that emerge through the modeling exercise (Frantz and Siddiki, 2022; Frantz, 2015; Ghorbani and Bravo, 2016; Smajgl, Izquierdo and Huigen, 2008). With the former, the institutional analyst may rely on IG coded statements to gain an understanding of the arrays of actors (i.e., agents) whose behavior should be reflected in one's model, the various actions associated with these actors, and action constraints and outcomes. With the latter, the institutional analyst first uses a modeling exercise to understand patterns in agent behaviors that emerge when agents are endowed with attributes and assigned different choice sets. The IG is then applied to characterize behavioral patterns in terms of institutional statements and corresponding syntactic components.

The noted usage of the IG in ABM is enabled by an intuitive mapping of the IG's syntactic components to key modeling parameters. Attributes information conveys information regarding agents to include in one's modeling exercise. Aim, Deontic, and Context information, used together, conveys the action sets associated with Attributes (agents) as well as associated discretion and consequences. Or else information conveys consequences associated with actions. Syntactic components of constitutive statements can also be instructive for model parameterization. Constituted Entities detail systemic artifacts on which, or through which, agents individually and collectively behave.

ABM has various affordances for IG analysis. First, ABM fundamentally operates on configural representations of institutional statements – realized in model design or interpreted in model results. It is thus ideally suited to understanding linkages among institutional statements and overcoming biases associated with viewing information associated with a particular institutional statement in isolation. Second, ABM enables 'institutional experimentation'; that is, the ability to observe types of behaviors that emerge within institutional constraints and selective manipulation of those.

The IG and ABM have been paired in the study of natural resource governance issues. Ghorbani and Bravo (2016) use the IG in both the up-front parameterization of an ABM, as well as to characterize emergent institutions in a common pool resource setting. Ghorbani and Bravo (2016) start their exercise with a set of institutional statements that follow the IG syntactic structure – delineating what an actor is required, allowed, or forbidden to do within a particular constraint. For example, 'Actor must appropriate n resource units at the time a condition is met'. Their study evaluates how these statements are adapted over time by actors endowed with characteristics and decision-making constraints. Siddiki and Frantz (2021) use IG-coded regulations governing the organic farming industry alongside survey and interview data to design an agent-based model that evaluates how farmer compliance changes under different institutional scenarios. IG coded regulation data were used to help define the action sets agents face in different regulatory situations. Survey and interview data were used to construct agent profiles and related expectations of how they behave.

Network analysis with IG data

Network analysis examines ties (edges) among actors and objects (nodes) in *one or more cases* and is used to explore patterns of interactions among nodes, as well as explore exogenous and endogenous factors that structure whole networks. The patterns of interactions are used to compare the roles or positions of actors in networks or the patterns of interactions across networks, as well as the association of differently structured networks with various outcomes (Schlager et al., 2022). Policy network analysis has a long tradition in studies of natural resource governance with survey data used to identify reported interactions among actors (Yi and Scholz, 2015). Typically, these studies focus on behavior, with little consideration given to how institutional arrangements may guide or influence reported behavior (see Olivier and Schlager (2021) for a review/discussion).

The grammar of institutions allows for institutional arrangements to be treated as networks of prescribed interactions and incorporated into network analyses (Olivier, 2019). To do so, analysts take Grammar-coded data and build matrices suitable for network analyses. The steps for doing so are carefully described in Herzog et al. (2022) and Olivier and Schlager (2021).

The Attributes and Object components of the Grammar are used to identify network nodes. Edges, or ties, among nodes may be rule types, Deontics, or Aims (see Olivier, Scott and Schlager (2020) for additional Grammar-based approaches for constructing networks). In identifying nodes and edges, it is important to maintain the context and meaning of the configuration of institutional statements being analyzed.

Descriptive, correlational, and inferential approaches may be used for analyses. Olivier and Schlager (2021) use all three approaches in explaining how actors' choices and institutional arrangements together structure collaborative activities. They construct a perceived network of interaction based on survey data, and a prescribed network of interaction based on Grammar-coded data. The perceived network of interaction was constructed from a survey of participants in the New York City Watersheds governing arrangement asking them to identify up to ten other participants with whom they interact. In addition, the participants were asked to identify up to five actors who they perceived as most influential in the watersheds, used to construct a reputation variable. The prescribed network of interaction was constructed from Grammar-coded data of the water quality and economic development programs, watershed rules and regulations, and the memorandum of agreement that created the governing arrangement. All regulative statements (ties) containing animate attributes and objects (nodes) were used to create the networks.

Descriptive measures capture different features of networks created by patterns of linkages. Measures such as density and centrality are common descriptive measures. Density captures all the linkages in a network divided by all possible linkages that could occur in a network. It is often used as a measure of bonding capital (Olivier and Schlager, 2021). Centrality, in general, captures the number of linkages needed to connect actors. It is often used as a measure of bridging capital. Quadratic Assignment Procedure (QAP) is a common approach for *comparing* networks based on correlations. 'QAP coefficients can be understood as the probability that if a linkage exists between actors a and b in network 1, that very same linkage is observed between the same actors in network 2' (Olivier and Schlager 2021: 354). Finally, an Exponential Random Graph Model (ERGM) is a type of logit regression model used to explore the effect of endogenous and exogenous (to the network) factors on actors' selection of partners in a network. 'A statistically significant coefficient indicates that the parameter of interest is significantly more (less) present in the observed network than in networks that could be generated by chance' (Olivier and Schlager 2021: 354). Olivier and Schlager (2021) incorporated the perceived network of interaction (based on survey data) and the prescribed network of interaction (based on Grammar-coded institutional statements) into an ERGM to examine the influence of features of both networks on the observed collaborative network. They find that aspects of the prescribed networks, in particular the power sharing decision venues, and aspects of the perceived network, for example, reputation, influence the structure of the collaboration.

A different approach using Grammar-coded data for network analyses is illustrated by Heikkila and Weible (2018). Instead of examining or comparing specific cases, or networks of ties among actors, they 'map' a policy subsystem and explore its polycentric features. Examining a policy subsystem requires the coding of dozens of laws, regulations, executive orders, and other types of policy documents. Manual coding is not feasible for these purposes and Heikkila and Weible (2018) propose a semi-automated approach. Using Automap, a type of text analysis software, they coded the eleven laws and regulations composing the US state of Colorado's oil and gas policy subsystem (Heikkila and Weible, 2018). Specifically, they coded the actors (Attributes and Objects), the Deontics, and rule types (based on Aims) of all statements composing each law and regulation. Using Ora and Ucinet, two network software programs, they examined rule and Deontic ties among the different actors prescribed to interact in the oil and gas laws and rules and explored the centrality of different actors across the laws and regulations.

Conclusion

Scholars are increasingly recognizing the value of the IG approach for analyzing natural resource governing arrangements. The number of publications that utilize the IG is rapidly growing, supported by an extensive global network of scholars as well as a rich set of resources that make the IG widely accessible (see the Institutional Grammar Research Initiative at institutionalgrammar.org). The development of the IG 2.0 by Frantz and Siddiki (2022) provides the foundation for extending its application to computational social science, both through machine learning for coding, and agent-based modeling for exploring the performance of a wide range of natural resource governance settings. The IG is a proven analytical approach for systematically analyzing natural resources institutional arrangements.

Some pressing theoretical and methodological issues remain that require attention. As the IG becomes increasingly accessible, and machine learning techniques for coding become more widely used, large sets of text-based data will become available, and standards for best practices will need to be developed. The meaning of the statements in context must be preserved. One approach for doing so is the creation of a glossary consisting of the operationalization and measurement of theoretically meaningful concepts and variables. In addition, institutional statements contain information on space (e.g., location and geography) and time, and institutional statements are revised over time. These dimensions have not been adequately explored. Frantz and Siddiki, with the IG 2.0, offer a basis for systematically capturing these dimensions in the coding process, but the approach they offer requires empirical validation for studying time and space. Finally, initial progress has been made in combining institutional arrangements and human behavior in modeling. Agent-based modeling and network analysis techniques seem especially promising for exploring the interaction of actors in various institutional settings (see Siddiki and Frantz 2021, Olivier and Schlager 2021, and Herzog et al., 202). More work in these areas promises to produce important theoretical breakthroughs on institutional design and performance.

References

Baggio, J., Barnett, A., Perez-Ibarra, I., Brady, U., Ratajczyk, E., Rollins, N., Rubinos, C., Shin, H., Yu, D., Aggarwal, R., Anderies, J.M. and Janssen, M. (2016) Explaining success and failure in the commons: The configural nature of Ostrom's Institutional design principles, *International Journal of the Commons*, 10(2): 417–439. https://doi.org/10.18352/ijc.634

Chen, C., Heikkila, T., Weible, C.M., Yordy, J., Yi, H., Berardo, R. and Kagan, J. (2022) Policy composition and adoption duration: Capturing conflict in the legislative process, *Policy Studies Journal*, 50(2): 407–431. https://doi.org/10.1111/psj.12457

Crawford, S. and Ostrom, E. (1995) A grammar of institutions, *American Political Science Review*, 89(3): 582–600. https://doi.org/10.2307/2082975

Clement, S., Moore, S.A. and Lockwood, M. (2015) Authority, responsibility and process in Australian biodiversity policy, *Environmental and Planning Law Journal*, 32(2): 93–114.

Frantz, C.K. (2015) *Agent-Based Institutional Modelling: Novel Techniques for Deriving Structure from Behaviour*, PhD thesis, Dunedin: University of Otago.

Frantz, C. and Siddiki, S. (2022) *Institutional Grammar: Foundations and Applications for Institutional Analysis*, Cham: Palgrave Macmillan. https://doi.org/10.1007/978-3-030-86372-2

Ghorbani, A. and Bravo, G. (2016) Managing the commons: A simple model of the emergence of institutions through collective action, *International Journal of the Commons*, 10(1): 200–219. https://doi.org/10.18352/ijc.606

Hanlon, J., Olivier, T. and Schlager, E. (2019) Suspicious collaborators: How governments in polycentric systems monitor behavior and enforce public good provision rules against one another, *International Journal of the Commons*, 13(2): 977–992. https://doi.org/10.5334/ijc.924

Heikkila, T. and Weible, C.M. (2018) A semiautomated approach to analyzing polycentricity, *Environmental Policy and Governance*, 28(4): 308–318. https://doi.org/10.1002/eet.1817

Herzog, L., Ingold, K. and Schlager, E. (2022) Prescribed by law and therefore realized? Analyzing rules and their implied actor interactions as networks, *Policy Studies Journal*, 50(2): 366–386. https://doi.org/10.1111/psj.12448

Kiser, L. and Ostrom, E. (1982) The three worlds of action: A metatheoretical synthesis of institutional approaches, in Ostrom E. (ed.) *Strategies of Political Inquiry*, Thousand Oaks, CA: Sage.

Leslie, H., Basurto, X., Nenadovic, M., Sievanen, L., Cavanaugh, K.C., Cota-Nieto, J.J., Erisman, B.E., Finkbeiner, E., Hinojosa-Arango, G., Moreno-Báez, M., Nagavarapu, S., Reddy, S.M.W., Sánchez-Rodríguez, A., Siegel, K., Ulibarria-Valenzuela, J.J., Weaver, A.H. and Aburto-Oropeza, O. (2015) Operationalizing the social-ecological systems framework to assess sustainability, *Proceedings of the National Academy of Sciences (PNAS)*, 112(19): 5979–5984. https://doi.org/10.1073/pnas.1414640112

McCord, P., Dell'Angelo, J., Baldwin, E. and Evans, T. (2016) Polycentric transformation in Kenyan water governance, *Policy Studies Journal*, 45(4): 633–658. https://doi.org/10.1111/psj.12168

McGinnis, M. (2011) Networks of adjacent action situations in polycentric governance, *Policy Studies Journal*, 39(1): 51–78. https://doi.org/10.1111/j.1541-0072.2010.00396.x

Novo, P. and Garrido, A. (2014) From policy design to implementation: an institutional analysis of the new Nicaraguan Water Law, *Water Policy*, 16(6): 1009–1030. https://doi.org/10.2166/wp.2014.188

Olivier, T. (2019) How do institutions address collective-action problems? Bridging and bonding in institutional design, *Political Research Quarterly*, 72(1): 162–176. https://doi.org/10.1177/1065912918784199

Olivier, T, Scott, T. and Schlager, E. (2020) Institutional design and complexity: Protocol network structure in response to different collective-action dilemmas, in Fischer, M. and Ingold, K. (eds.) *Networks in Water Governance*, Cham: Palgrave Macmillan, pp. 267–293.

Olivier, T. and Schlager, E. (2021) Rules and the ruled: Understanding joint patterns of institutional design and behavior in complex governing arrangements, *Policy Studies Journal*, 50(2): 340–365. https://doi.org/10.1111/psj.12429

Ostrom, E. (1990) *Governing the Commons*, Cambridge: Cambridge University Press.

Ostrom, E. (1999) Coping with the tragedies of the commons, *Annual Review of Political Science*, 2: 493–535. https://doi.org/10.1146/annurev.polisci.2.1.493

Ostrom, E. (2005) *Understanding Institutional Diversity*, Princeton, NJ: Princeton University Press.

Ostrom, E. (2009) A general framework for analyzing sustainability of social-ecological systems, *Science*, 325(5939): 419–422. https://doi.org/10.1126/science.1172133

Ostrom, E., Agrawal, A., Blomquist, W., Schlager, E. and Tang Y. (1989) *The CPR Coding Manual*, Bloomington: Workshop in Political Theory and Policy Analysis, Indiana University.

Ostrom, E., Gardner, R. and Walker, J. (1994) *Rules, Games, and Common Pool Resources*, Ann Arbor: University of Michigan Press.

Ostrom, E. and Basurto, X. (2011) Crafting analytical tools to study institutional change, *Journal of Institutional Economics*, 7(3): 317–343. https://doi.org/10.1017/S1744137410000305

Ostrom, V. and Ostrom, E. (1977) Public goods and public choices, in Savas, E.S. (ed.) *Alternatives for Delivering Public Services: Toward Improved Performance*, Boulder, CO: Westview Press.

Pieper, L., Ruiz, S., Schlager, E. and Schweik, C. (2023) The use of the institutional grammar 1.0 for policy analysis: A literature review, *International Journal of the Commons*, 17(1): 256–270. https://doi.org/10.5334/ijc.1214

Poteete, A., Janssen, M. and Ostrom, E. (2010) *Working Together: Collective Action, the Commons, and Multiple Methods in Practice*, Princeton, NJ: Princeton University Press.

Schlager, E. and Heikkila, T. (2011) Left high and dry? Climate change, common pool resource theory, and the adaptability of western water compacts, *Public Administration Review*, 71(3): 461–470. https://doi.org/10.1111/j.1540-6210.2011.02367.x

Schlager, E. and Ostrom, E. (1990) Property rights regimes and natural resources: A conceptual analysis, *Land Economics*, 68(3): 249–262. https://doi.org/10.2307/3146375

Schlager, E. and Ostrom, E. (1992) Property rights regimes and natural resources, *Land Economics*, 68(3): 249–269. https://doi.org/10.2307/3146375

Schlager, E., Sidikki, S. and Cox, M. (2022) Methods for analyzing social dilemmas and institutional arrangements within the institutional analysis and development framework, in Weible, C.M. and Workman, S. (eds.) *Methods of the Policy Process*, New York: Routledge. pp. 208–232.

Schlager, E. and Villamayor Tomas, S. (2023) The IAD framework and its tools for policy and institutional analysis, in Weible, C. (ed.) *Theories of the Policy Process*, 5th edition, New York: Routledge, pp. 196–229.

Schlager, E.C., Bakkensen, L.A., Olivier, T. and Hanlon, J. (2021) Institutional design for a complex commons: Variations in the design of credible commitments and the provision of public goods, *Public Administration*, 99(2): 263–289. https://doi.org/10.1111/padm.12715

Siddiki, S. and Frantz, C.K. (2021) New opportunities for institutional analysis in public administration research, *Public Administration*, 99(2): 213–221. https://doi.org/10.1111/padm.12753

Siddiki, S. and Frantz, C.K. (2023) Understanding the effects of social value orientations in shaping regulatory outcomes through agent-based modeling: An application in organic farming, *International Review of Public Policy*, 5(2): 203–235. https://doi.org/10.4000/irpp.3398

Siddiki, S., Heikkila, T., Weible, C.M., Pacheco-Vega, R., Carter, D., Curley, C., Deslatte, A. and Bennett, A. (2022). Institutional analysis with the institutional grammar, *Policy Studies Journal*, 50(2): 315–339. https://doi.org/10.1111/psj.12361

Smajgl, A., Izquierdo, L.R. and Huigen, M. (2008) Modeling endogenous rule changes in an institutional context: The ADICO sequence, *Advances in Complex Systems*, 11(2): 199–215. https://doi.org/10.1142/S021952590800157X

Watkins, C., and Westphal, L.M. (2015) People don't talk in institutional statements: A methodological case study of the institutional analysis and development framework, *Policy Studies Journal*, 44(S1): S98–S122. https://doi.org/10.1111/psj.12139

Yi, H. and Scholz, J.T. (2015) Policy networks in complex governance subsystems, *Policy Studies Journal*, 44(3): 248–279. https://doi.org/10.1111/psj.12141

4 Measuring institutions in quantitative research

Graham Epstein

Introduction

Institutions, which are the rules, norms and shared strategies (Crawford and Ostrom, 1995; McGinnis, 2011a) that groups use to structure their interactions with the environment and each other, play a critically important role in the pursuit of environmental sustainability (Bayu, Kim and Oki, 2020; Cinner et al., 2016; Sze et al., 2022). Marine protected areas, for instance, tend to be more ecologically effective when fishers are engaged in the development of rules that affect them (Di Franco et al., 2016) and those rules are enforced (Di Franco et al., 2016; Edgar et al., 2014). Indeed, there is mounting empirical evidence that local participation in rulemaking (Chhatre and Agrawal, 2009; Persha, Agrawal and Chhatre, 2011) and local enforcement (Chhatre and Agrawal, 2008; Rustagi, Engel and Kosfeld, 2010) provide an important foundation for collective action and sustainable governance of natural resources (Baggio et al., 2016; Cox, Arnold and Villamayor Tomas, 2010; Epstein, 2017; Ostrom, 1990). Nonetheless, there are also many cases where legislative changes that seem to enable local participation in rulemaking have failed to yield meaningful changes in environmental conditions (Adjei, Agyei and Adjei, 2020; Blaikie, 2006; Coleman and Fleischman, 2012; Ocampo-Diaz et al., 2022; Ribot, Agrawal and Larson, 2006). While on the one hand these conflicting results could stem from differences in sectors, contexts, measures of 'success' or simply a lack of adequate time for institutions to have environmental effects, they could also relate to differences in how institutions are measured across different studies. Indeed, although there have been longstanding efforts to promote consistent use of language, concepts and measures by scholars associated with the Bloomington School of institutional analysis (Crawford and Ostrom, 1995; McGinnis, 2011a; Ostrom, 2007; Poteete, Janssen and Ostrom, 2010), this goal remains somewhat elusive. This chapter, therefore, adapts logic models, a tool that is frequently used to support program design and evaluation (Bucher, 2010; Knowlton and Phillips, 2012; McLaughlin and Jordan, 2015) to develop a typology of five major types or approaches for measuring institutions.

DOI: 10.4324/9781032624679-4

Before turning to these different approaches or types of institutional measures, it is important to clarify how I and many researchers associated with the Bloomington School of Institutional Analysis define institutions. Institutions are human-constructed templates or statements that specify actions that must, must not or may be taken by actors and the conditions in which those statements apply (Crawford and Ostrom, 1995). In general, there are three major types of institutions, including formal and informal rules, social norms and conventions or shared strategies. Formal and informal rules are defined by the presence of sanctions for individuals that fail to adhere to them, and associated systems of enforcement. Formal rules (or rules-in-form, *de jure* rules) are those rules that are written down in management plans, policies and legislations, while informal rules (or rules-in-use, *de facto* rules) are those that are actually used by a group of relevant actors to govern their interactions and generally exist as a shared understanding in the minds of a community or group (Ostrom, 2005). Social norms, meanwhile, are similar to rules in that they specify actions that actors must, must not or may take, and the conditions in which those social norms apply, but differ in that social norms lack clearly specified sanctions and enforcement systems for sustaining them. Instead, social norms are generally upheld by intrinsic incentives and/or social sanctions (Andreoni, 1989; Bowles and Gintis, 2009; Coleman, 2003). Finally, conventions or shared strategies are simply statements that specify actions to be taken by individuals under certain conditions and are typically used to resolve coordination problems such as how to move efficiently in a crowded space or road (i.e., keep right or keep left).

Institutional researchers also often distinguish between operational rules and collective choice rules. Operational rules are used to govern a range of practical activities, including the harvesting, processing and sale of resources, and management actions such as maintenance, monitoring, sanctioning and conflict resolution. Operational rules can specify where, when and how individuals are permitted to harvest resources, including restrictions on the technologies they can use (i.e., minimum mesh sizes) and the types of resources they are allowed to retain (i.e., restrictions on certain species, or the size and/or sex of harvested resources). Measures based on operational rules can in turn be used to inform research that seek to develop a better understanding of the relationship between harvest control rules and a range of potentially relevant social and ecological outcomes. For example, researchers have found that individual and community quotas (a form of quantity rule) tend to be associated with more sustainable fisheries (Costello, Gaines and Lynham, 2008; Gutierrez, Hilborn and Defeo, 2011), while others have linked patterns of deforestation to protected areas (Barber et al., 2014; Rahman and Islam, 2021). However, operational rules also include rules specifying how actors undertake governance activities such as who is permitted to monitor compliance with rules.

Collective choice rules, meanwhile, define how operational rules are made, including who is eligible to participate in decision-making processes and how decisions are made. In general, collective choice rules define to a great extent the power that different groups hold concerning the use and management of resources and thus measures of collective choice rules tend to focus on themes of representation and authority. For example, several studies have linked stakeholder participation in decision-making processes to higher levels of compliance and collective action in natural resource management (Alexander et al., 2018; Viteri and Chávez, 2007), while others have revealed similar findings with respect to measures that indicate whether rules were designed by local stakeholders (Epstein, 2017; Epstein et al., 2021). Importantly, however, the relationship between participation rights defined by legislation or policies, collective action and sustainability is far less clear (Adjei, Agyei and Adjei, 2020; Blaikie, 2006; Coleman and Fleischman, 2012; Ocampo-Diaz et al., 2022; Ribot, Agrawal and Larson, 2006). Thus, the types of measures that are used to represent high-level concepts such as participation and assess their relationship with outcomes of interest can have important implications for institutional theory and the design and implementation of public policies.

Measuring institutions

Institutional analysis of SESs has benefited greatly by the development of frameworks that help to structure research on the sustainability of SESs, and the role of institutions in these systems. Indeed, the institutional analysis and development framework (Ostrom, 2005), social–ecological systems framework (Ostrom 2009) networks of action situations (Cole, Epstein and McGinnis, 2019; McGinnis, 2011b) and grammar of institutions (Crawford and Ostrom, 1995; Siddiki et al., 2022) are among the most widely used frameworks for organizing research and identifying variables and processes that influence sustainability. However, despite some advances with the institutional grammar, researchers have struggled to achieve consistency with respect to the measures they use to operationalize these variables, including institutions (Cumming et al., 2020; Partelow, 2018). This measurement gap likely stems from at least three core challenges.

First, institutional analysis in an interdisciplinary field of research with researchers applying a wide range of research methods, ranging from participant observation, interviews and surveys to laboratory experiments and mathematical models (Poteete, Janssen and Ostrom, 2010). Second, researchers develop measures of institutions for different purposes that influence how they are measured. Researchers interested in advancing understanding of the relationship between institutions, sustainability or well-being will tend to employ analytical measures that record details about membership in different classes (i.e., state or community, high or low participation), or levels thereof (i.e., number of guards hired to conduct patrols, percentage of households reporting participation in meetings). In contrast, researchers interested in institutional evaluations will tend to begin by defining attributes of an 'effective' or 'successful' institution that is distinct from the broader social and environmental objectives. For example, evaluative indicators may record whether management plans establish mechanisms for conflict resolution, or alternatively the percentage of stakeholders reporting satisfaction with or acceptance of rules. While this approach is often useful to uncover potential institutional strengths and weaknesses, the utility of these indicators ultimately rests upon the assumptions that were made regarding the characteristics of effective institutions. Furthermore, in cases where subjective measures are used to form evaluations, it can be difficult to separate subjective assessments of institutions from perceptions of their social and ecological impacts. Third and finally, because the institutions that are established on paper through legislation and management plans are often incomplete, or deviate substantially from the institutions that are used in practice (Helmke and Levitsky, 2004; Maynard-Moody and Portillo, 2010), measures based on these formal institutions may be misleading and/or exhibit considerable variability across a population of cases. As a result, many institutional researchers move beyond simply measuring rules, to collect details about the decisions that actors take to implement institutions, and the outcomes of those decisions.

Logic models are frequently used in applied policy settings as a tool for outlining a theory of change for programs by building a causal chain that connects policy inputs to outcomes through one or more activities and outputs. Logic models are often developed in the planning or design stage of programs, providing a tool for critically reflecting on the efficacy of achieving desired program outcomes (McLaughlin and Jordan, 2015). In an ideal situation, logic models begin by specifying the desired outcomes of a program and work in reverse to determine the outputs, activities and inputs that are needed to achieve these outcomes. However, in practice, analysts will typically work in both directions, adjusting outcomes to better reflect the inputs that are available. Once a logic model is completed, program staff will typically specify indicators for each stage of the causal chain and use these to evaluate the overall effectiveness of the program. While policy content is not typically included in applied logic models where programs are typically designed in the context of pre-defined policies, they are a particularly important type of indicator for institutional research. As a result, the adapted logic model presented in Figure 4.1 includes five major classes or types of measures used in institutional research. These include indicators derived from (i) assessments of the content of formal and informal rules, (ii) the inputs or resources that groups allocate to support governance activities, (iii) the activities undertaken by governance actors, (iv) the outputs of those activities in terms of the goods and services that are produced, and (v) the outcomes or effectiveness of those institutions in achieving their intended functions.

The following sections of this chapter describe how indicators of institutions are developed as researchers move from institutional content specified in legislation, management plans or collective understanding of communities to generate institutional outcomes related to their functions. This is then followed by a summary in the "Synthesis" section.

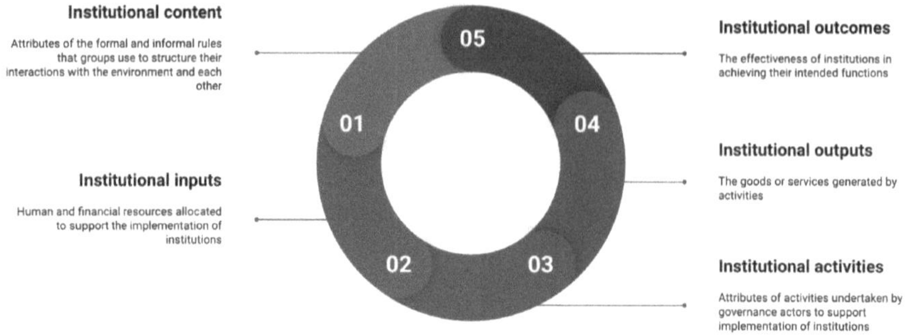

Figure 4.1 Approaches to measuring institutions in institutional research.

Institutional content

The content of rules, outlined in formal government documents, such as international agreements, government legislation, regulations, policies and management plans, or informal rules elicited through surveys, interviews or focus group discussions, unsurprisingly play a foundational role in research on institutions. At a high level, content-based measures of institutions provide insights about the types of rules, or attributes thereof, that groups use to structure their interactions with the environment and with each other. Rule typologies or other coding systems that have been developed over the years are often helpful in this regard. For example, operational rules that govern the use of resources often distinguish between quantity (i.e., quotas), timing (i.e., closed and open seasons, harvest intervals), location (protected areas, closures or other spatial restrictions), character (i.e., size or species restrictions) and technology (i.e., gear restrictions or requirements) rules (Acheson, Wilson and Steneck, 1998; Schlager, Blomquist and Tang, 1994). Property rights, on the other hand, can be measured based on the bundle of access, withdrawal, management, exclusion and alienation rights that groups possess in relation to a particular resource (Miller et al., 2021; Schlager and Ostrom, 1992). Researchers may also consider using the increasingly formalized grammar of institutions for coding the elements of regulatory statements (Crawford and Ostrom, 1995; Frantz and Siddiki, 2021; Siddiki et al., 2022).

Content-based measures of institutions derived from secondary data generally involve systematic coding of rules or the attributes thereof found within regulatory documents like legislation or management plans. For example, Blackman et al. (2017) drew upon a dataset of land records to find that areas titled (i.e., a bundle of property rights) to Indigenous communities in the Peruvian Amazon experienced least short-term declines in land clearing and disturbance. Tilleard and Ford (2016), meanwhile, assessed the adaptive capacity of transboundary river basins by reviewing national-level legislation and calculating the percentage of states associated with that river basin that have a statutory requirement for adaptation. In contrast, measures of informal rules are often elicited through a range of primary research methods, including surveys, interviews, focus groups and participant observation, or triangulation across multiple methods. The International Forest Resources and Institutions (IFRI) program for instance, asks researchers to integrate insights from multiple methods to assess whether members of a local user group are responsible for making *de facto* rules for the use of local resources (Persha, Agrawal and Chhatre, 2011).

Widespread use of content-based measures in institutional research stems, at least in part, from the accessibility of source material (i.e., laws, policies and management plans) that enables relatively low-cost desk-based research. However, content-based measures tend to suffer from at least one of two core challenges. First, desk-based coding of policy content often involves significant challenges in ensuring consistency across different policy documents, coders and studies (Bushouse et al., 2023; Frantz and Siddiki, 2021). Second, even if policies are coded relatively consistently, the implementation of those policies can vary widely across different cases. Indeed, paper parks are fairly common in biodiversity conservation where the seemingly strict rules on paper belie the fact that those rules are not enforced, and sometimes not even known by local stakeholders (Bustamante et al., 2014; Claudet, 2018; Relano and Pauly, 2023).

Institutional inputs

The resources that governments, communities and other stakeholders allocate towards the implementation of institutions are another way that institutional researchers measure institutions in quantitative research. In general, input-based measures are based on the resources that actors, including governments, non-governmental organizations and communities allocate to support the implementation of formal or informal institutions. Although a number of different types of inputs may be relevant for different types of institutions or contexts, input-based measures are generally derived from information about the human and/or financial resources allocated to particular activities or tasks. Agrawal and Goyal (2001), for instance, measured the strength of local enforcement systems on the basis of revenues raised from the community for local enforcement. Indeed, adequate budgets are frequently highlighted as a key ingredient in sustainable natural resource management and conservation (Gill et al., 2017; Rife et al., 2013), providing a foundation for many critically important, but costly governance activities such as environmental monitoring and enforcement. In contrast, Scott (2015) assessed formalization of collaborative watershed organizations based on the human resources these organizations possessed to support planning and implementation of governance activities. This research found that the ecological effectiveness of collaborative watershed organizations tended to increase when those organizations had dedicated coordinators.

Much like content-based measures, input-based measures are often based on records developed and maintained by governments or other relevant organizations that provide details about staffing and financial allocations for the implementation of institutions. Rights to participate in decision-making processes rules may, for instance, have limited meaningful impact if managers lack staff to engage with local communities and/or financial resources to organize and hold meetings with those communities (Coad et al., 2019; Cumming et al., 2022; Gill et al., 2017). Input-based measures thus provide some indication of the extent to which institutions expressed on paper are translated into resources for the implementation of those institutions. However, input-based measures also tend to suffer from a number of different limitations that may limit their utility in research. First, budgets or staffing required to achieve given social and ecological objectives can vary widely across different social and ecological contexts. Agrawal and Goyal (2001), for example, describe how the costs of effective third-party enforcement systems rise rapidly with increases in the size of communities and spatial distribution of resources. As a result, raw budgets or staffing can be misleading and often need to be weighted by relevant features of the social and ecological context, such as population or area, to enable better comparisons of institutions. However, it can be difficult to determine which of the many features of complex SESs (Agrawal, 2003; Ostrom, 2009) are relevant for weighting. Second, although inputs can clearly enable implementation of institutions, these resources are not always allocated effectively to undertake activities to generate outputs and successful outcomes. Groups may, for instance, have staff to monitor compliance with rules, but lack equipment such as patrol boats in working condition to undertake monitoring activities (Catedrilla et al., 2012; Obiero et al., 2015)

Institutional activities

Activity-based measures of institutions provide details about the processes undertaken by governance actors to implement institutions, providing a critically important and tangible link between policy content and inputs to desired outcomes. These indicators provide insights into the functioning of governance system, based on measures of relevant activities such as the number of public meetings (participation), patrols (enforcement), ecological assessments (ecological monitoring) and adjudications (conflict resolution). Activity-based measures can be derived from a range of different sources, including a review of public or organizational records, surveys, interviews and participant observation. Epstein (2017), for example, drew upon data collected as part of the IFRI program that combines insights from participant observation and interviews to code details of about local institutional arrangements for forest governance, including local enforcement. The local enforcement indicator, which has been used in several studies using IFRI data (Chhatre and Agrawal, 2008; Coleman, 2009; Epstein et al., 2021; Gibson, Williams and Ostrom, 2005), measures enforcement based on the relative frequency that members of local user groups self-organize to monitor compliance with rules.

Activity-based measures are frequently used in academic and applied research on institutions and have several important strengths when compared to alternative approaches. First, they are often favoured by applied researchers by providing leading indicators of potential problems due to gaps in terms of activities such as public consultations, enforcement patrols or ecological monitoring. Second, activities can be particularly useful for measuring informal institutions (or rules-in-use) where rules exist as a collective understanding of a community or group. In some cases, it can be difficult to elicit information about informal institutions that may be deeply institutionalized and taken for granted by members of a community. As a result, observing patterns of activities and/or developing instruments to record details of relevant activities such as patrols, public meetings and sanctions can be particularly useful for capturing details about informal institutions. Ostrom (1990), for example, inferred that enforcement was produced as a by-product of appropriation in certain irrigation systems where timing rules ensured that irrigators could observe the behaviour of the actors that preceded them. However, activity-based measures struggle with many of the same challenges as input-based measures in that the level or quality of activities required to achieve desired outcomes can vary widely across different social and ecological contexts (Harden-Davies et al., 2022). More public meetings where participation is limited to receiving information from governance actors may, for instance, prove less effective than fewer meetings where resource users are able to play a more active role in designing or adapting the rules that affect them (Cox, Arnold and Villamayor Tomas, 2010).

Institutional outputs

Output-based measures of institutions are closely related to activity-based measures but shift their attention to measuring the immediate results or products generated by those activities. Indeed, output-based measures begin to shift their attention away from the decisions or actions of governance actors and towards the resource users and other stakeholders that are subject to those actions or decisions. For example, self-reported participation in meetings or decision-making processes is often used as an indicator for collective-choice rules. At the individual level, these indicators can be used to explore important theoretical questions concerning the impacts of participation on a range of desirable behaviours and outcomes, including rule compliance (Viteri and Chávez, 2007), social enforcement of conservation rules (Alexander et al., 2018), livelihoods and resource conditions (Cinner et al., 2012). They can also easily be aggregated for group-level reporting and analysis by calculating the number of participants in public meetings or fraction of respondents in surveys reporting attendance at relevant meetings or decision-making processes. Fidler et al. (2022), for instance, used a household survey to collect information about participation in meetings at several stages of the decision-making process for marine protected areas and found that participation was generally associated with increases in fish biomass over time.

Output-based measures of institutions are fairly common in applied research and program evaluation where donors and government agencies often seek to understand and evaluate the immediate outcomes of their investments (Howe and Milner-Gulland, 2012). They may, for instance, be interested in understanding how many stakeholders attended participatory planning or capacity building sessions, or alternatively how many rule violations were detected as a result of investments in human resources or equipment for enforcement. However, with the notable exception of participation in collective-choice processes, output-based measures are rarely used in academic research on institutions. Indeed, the lack of clearly defined thresholds for good or adequate outputs and wide variance in social and ecological contexts can make it difficult to establish useful output-based indicators for cross-case comparison. Furthermore, attendance at meetings or capacity building sessions does not necessarily mean that actors had any meaningful influence over decisions (Sen and Nielsen, 1996), or acquired relevant skills or knowledge to build their capacity. As a result, many experts have urged a shift away from activity and output-based reporting to focus on outcomes for assessing institutions and the impacts of institutional change (Bottrill, Hockings and Possingham, 2011; Harden-Davies et al., 2022; Howe and Milner-Gulland, 2012).

Institutional outcomes

Outcome-based measures are the fifth and final way in which institutions are measured in applied and academic research and are generally based on an assessment of the extent to which rules or policies are effective in achieving their intended or desired functions. More specifically, outcomes do not refer to the broader social or environmental objectives of institutions, such as the condition or abundance of resources, but rather to the functional contributions of those institutions. For example, institutions that enable stakeholder participation are generally used to (i) help improve the fit between operational rules and local conditions (Chhatre and Agrawal, 2009; Drazkiewicz, Challies and Newig, 2015), (ii) enhance the legitimacy and acceptance of rules (Nielsen, 2003; Trachtenberg and Focht, 2005), and/or (iii) satisfy innate psychological needs for self-determination (DeCaro, Janssen and Lee, 2015; DeCaro and Stokes, 2013; Frey and Jegen, 2001; Ryan and Deci, 2000). Thus, the outcomes of participatory institutions may be assessed using a variety of different measures or combinations thereof. Researchers interested in understanding the fit between operational rules and local conditions might, for instance, assess the extent to which management boundaries are aligned with the boundaries of resources (Moss, 2012; Qiu et al., 2017). In contrast, researchers interested in understanding the impacts of participation on the legitimacy of those rules might ask questions concerning the perceived fairness and/or support for those rules (Ramcilovic-Suominen and Epstein, 2015).

In general, outcome-based measures are developed through one of three approaches. First as noted in the example above, researchers might define a desired outcome of an institution, such as stakeholder support for rules and develop survey questions to assess the extent to which rules are supported within a community of resource users. Second, many qualitative researchers use triangulation across multiple methods, including document review, surveys, interviews, focus groups and participant observation to develop subjective assessments of the effectiveness or outcomes of institutions (Orr et al., 2016; Rajkumar, Narayan and Mohammed Irshad, 2020). Indeed, triangulation offers a valuable approach for enhancing the accuracy of measures by combining multiple sources of information and knowledge systems (Cochrane et al., 2019; Thiault et al., 2021). However, this approach often relies upon subjective assessment, which can create challenges for cross-case comparison and can require significant investments of time and resources. As a result, researchers often turn to a third, lower cost, approach that draws upon the expertise and knowledge of key informants in interviews or focus groups to assess institutional outcomes (Johns, 2021; Mahajan et al., 2024; OECD, 2018; Stolton et al., 2019; UNEP, 2021). While this approach can help to offer relatively rapid and lower cost assessments, they can still struggle to generate comparable data. Furthermore, although at least some outcome-based measures hold considerable utility for applied evaluative and theoretical research, their utility for larger-n applied analytical research is somewhat limited. For example, if a research study finds that high levels of acceptance of rules (one potential indicator of effective stakeholder participation) are associated with more sustainable outcomes, it provides limited guidance regarding the policies or activities that might be needed to generate that acceptance.

Synthesis

This chapter set out to develop a tool for developing a better understanding of the ways in which researchers develop quantitative measures of institutions in their research and discuss the ways in which these choices serve to both enable and hinder advances in the theory and practice of sustainable environmental governance. The "Measuring institutions" section began by first highlighting the diversity of methods used to collect data for developing institutional measures, before continuing to discuss how the purpose of those measures can contribute to shaping them. More specifically, evaluative indicators tend to focus more on the outcomes of institutions (OECD, 2018) or alternatively require researchers to define thresholds or attributes of 'good' institutions, such as the alignment between the boundaries of a resource and its governance system (Milman et al., 2013). It then continued by introducing and adapting the concept of logic models to organize different measures of institutions as researchers move from reviewing the content of rules, to assessing allocations of resources, measuring institutional activities and outputs, and finally assessing the extent to which institutions achieve their intended functions, as outlined in Table 4.1.

Table 4.1 Measuring institutions in quantitative research

Diagnostic question	*Potential measures for participation*	*Potential measures for monitoring*	*Limitations*
Institutional content What are the rules or attributes thereof?	Presence of mechanisms for stakeholder participation in management plans	Presence of mechanisms designating rights to monitor compliance with rules	Differences between rules-in-form and rules-in-use Lack of implementation Rules are incomplete
Institutional inputs What resources have been allocated to implement institutions?	Budget for stakeholder participation	Number of monitors	Input measures may require standardization Effectiveness does not necessarily co-vary with inputs Inputs may not be relevant for measuring some institutions
Institutional activities What activities have been undertaken to implement institutions?	Number of public/community meetings	Number of person-hours for patrols	Activity-based measures may require standardization Effectiveness does not necessarily co-vary with activities Activities may not be relevant for measuring some institutions
Institutional outputs What goods or services have been produced as a result of activities?	Percentage of households reporting attendance at meetings	Number of offenses detected	Output-based measures may require standardization Effectiveness does not necessarily co-vary with outputs Outputs may not be relevant for measuring some institutions
Institutional outcomes Has the institution achieved its intended objective(s)?	Percentage of households expressing support for rules	Percentage of households reporting compliance with rules	Choice of objectives for measuring outcomes can influence results. High levels of reliance upon subjective measures Challenges in separating perceptions of institutional effectiveness from overall results

Table 4.1 synthesizes features of the five major classes of quantitative measures of institutions discussed in this chapter, including details concerning the high-level diagnostic questions they ask, examples for operationalizing these for assessing participatory and social monitoring institutions and brief summary of their limitations. For example, scholars interested in measuring institutions based on the activities undertaken by governance actors could develop a measure of institutions for participation and social monitoring based on the number of community meetings (Agrawal and Yadama, 1997), and number of person-hours of patrols. While activity-based measures help to address some of the limitations of content-based measures where institutions that are written on paper are either incomplete or entirely misleading (Bustamante et al., 2014; Relano and Pauly, 2023), measures of inputs, activities and outputs have their own, fairly similar set of limitations. First, not all institutions are clearly associated with particular inputs, activities or outputs. Most operational rules, for instance, specify the conditions in which resource users are permitted to harvest resources, which does not necessarily require governance actors to invest their time and resources into activities. Second, it can be very difficult to compare inputs, activities and outputs across observations with vast differences in terms of their social and ecological context, and thus some type of standardization is often required (Agrawal and Goyal, 2001). Third and finally, the effectiveness of institutions in achieving their intended functions does not necessarily co-vary with measures of inputs, activities or outputs. Larger budgets for stakeholder participation, and more meetings with higher participation rates may, for instance, have limited impacts on acceptance of rules and compliance if stakeholders have little meaningful influence in these meetings (Armitage et al., 2020; Frey and Jegen, 2001; Ruano-Chamorro, Gurney and Cinner, 2022). However, the solution that is often advanced to address this problem involving the measurement of outcomes has its own set of limitations. These include high levels of reliance upon the subjective perceptions of researchers or stakeholders and difficulties in separating perceptions of social and ecological outcomes from perceptions of the functional effectiveness of institutions. As a result, outcome-based measures of institutions can be difficult to compare across different cases and are often simply reflections of the broader social and ecological outcomes of those cases.

Institutional researchers have developed and utilized a wide range of different approaches for measuring institutions in their research. This diversity has enabled considerable advances in our understanding of the relationship between institutions, collective action and environmental sustainability (i.e., Ostrom, 1990). However, diverse measures can also confound efforts to synthesize evidence about these relationships if responses differ across measures of policy content, inputs, activities, outputs and outcomes. As a result, it is critically important that institutional researchers clearly define the types of measures they are using and are attentive to its limitations to support advances in institutional theory and the design of sustainable environmental governance systems.

References

Acheson, J.M., Wilson, J.A. and Steneck, R.S. (1998) Managing chaotic fisheries, in Berkes, F. and Folke, C. (eds.) *Linking Social and Ecological Systems*, New York: Cambridge University Press, pp. 390–413.

Adjei, P.O.-W., Agyei, F.K. and Adjei, J.O. (2020) Decentralized forest governance and community representation outcomes: Analysis of the modified taungya system in Ghana, *Environment, Development and Sustainability*, 22: 1187–1209. https://doi.org/10.1007/s10668-018-0243-7

Agrawal, A. (2003) Sustainable governance of common-pool resources: Context, methods, and politics, *Annual Review of Anthropology*, 32: 243–262. https://doi.org/10.1146/annurev.anthro.32.061002.093112

Agrawal, A. and Goyal, S. (2001) Group size and collective action: Third-party monitoring in common-pool resources, *Comparative Political Studies*, 34(1): 63–93. https://doi.org/10.1177/0010414001034001003

Agrawal, A. and Yadama, G. (1997) How do local institutions mediate market and population pressures on resources? Forest panchayats in Kumaon, India, *Development and Change*, 28(3): 435–465. https://doi.org/10.1111/1467-7660.00050

Alexander, S.M., Epstein, G., Bodin, Ö., Armitage, D. and Campbell, D. (2018) Participation in planning and social networks increase social monitoring in community-based conservation, *Conservation Letters*, 11(5): e12562. https://doi.org/10.1111/conl.12562

Andreoni, J. (1989) Giving with impure altruism: Applications to charity and Ricardian equivalence, *The Journal of Political Economy*, 97(6): 1447–1458. Available at https://www.jstor.org/stable/1833247

Armitage, D., Mbatha, P., Muhl, E.K., Rice, W. and Sowman, M. (2020) Governance principles for community-centered conservation in the post-2020 global biodiversity framework, *Conservation Science and Practice*, 2(2): e160. https://doi.org/10.1111/csp2.160

Baggio, J., Barnett, A., Perez-Ibarra, I., Brady, U., Ratajczyk, E., Rollins, N., Rubiños, C., Shin, H., Yu, D. and Aggarwal, R. (2016) Explaining success and failure in the commons: The configural nature of Ostrom's institutional design principles, *International Journal of the Commons*, 10(2): 417–439. https://doi.org/10.18352/ijc.634

Barber, C.P., Cochrane, M.A., Souza Jr, C.M. and Laurance, W.F. (2014) Roads, deforestation, and the mitigating effect of protected areas in the Amazon, *Biological Conservation,* 177: 203–209. https://doi.org/10.1016/j.biocon.2014.07.004

Bayu, T., Kim, H. and Oki, T. (2020) Water governance contribution to water and sanitation access equality in developing countries, *Water Resources Research*, 56(4): e2019WR025330. https://doi.org/10.1029/2019WR025330

Blackman, A., Corral, L., Lima, E.S. and Asner, G.P. (2017) Titling indigenous communities protects forests in the Peruvian Amazon, *Proceedings of the National Academy of Sciences*, 114(16): 4123–4128. https://doi.org/10.1073/pnas.1603290114

Blaikie, P. (2006) Is small really beautiful? Community-based natural resource management in Malawi and Botswana, *World Development*, 34(11): 1942–1957. https://doi.org/10.1016/j.worlddev.2005.11.023

Bottrill, M.C., Hockings, M. and Possingham, H.P. (2011) In pursuit of knowledge: Addressing barriers to effective conservation evaluation, *Ecology and Society*, 16(2): 14. https://www.ecologyandsociety.org/vol16/iss2/art14/

Bowles, S. and Gintis, H. (2009) Social preferences, Homo Economicus, and Zoon Politikon, in Buttenwieser, J.L. (ed.) *The Oxford Handbook of Contextual Political Analysis*, New York: Oxford University Press, pp. 172–186.

Bucher, J.A. (2010) Using the logic model for planning and evaluation: Examples for new users, *Home Health Care Management & Practice*, 22(5): 325–333. https://doi.org/10.1177/1084822309353154

Bushouse, B.K., Schweik, C.M., Siddiki, S., Rice, D. and Wolfson, I. (2023) The institutional grammar: A method for coding institutions and its potential for advancing third sector research, *Voluntas: International Journal of Voluntary and Nonprofit Organizations*, 34: 76–83. https://doi.org/10.1007/s11266-021-00423-w

Bustamante, G., Canals, P., Di Carlo, G., Gomei, M., Romani, M., Souan, H. and Vanzella-Khouri, A. (2014) Marine protected areas management in the Caribbean and Mediterranean seas: Making them more than paper parks, *Aquatic Conservation: Marine and Freshwater Ecosystems*, 24(S2): 153–165. https://doi.org/10.1002/aqc.2503

Catedrilla, L.C., Espectato, L.N., Serofia, G.D. and Jimenez, C.N. (2012) Fisheries law enforcement and compliance in District 1, Iloilo Province, Philippines, *Ocean & Coastal Management*, 60: 31–37. https://doi.org/10.1016/j.ocecoaman.2012.01.003

Chhatre, A. and Agrawal, A. (2008) Forest Commons and Local Enforcement, *Proceedings of the National Academy of Sciences*, 105(36): 13286–13291. https://doi.org/10.1073/pnas.0803399105

Chhatre, A. and Agrawal, A. (2009) Trade-offs and synergies between carbon storage and livelihood benefits from forest commons, *Proceedings of the National Academy of Sciences*, 106(42): 17667–17670. https://doi.org/10.1073/pnas.0905308106

Cinner, J.E., Huchery, C., MacNeil, M.A., et al. (2016) Bright spots among the world's coral reefs, *Nature*, 535: 416–419. https://doi.org/10.1038/nature18607

Cinner, J.E., McClanahan, T.R., MacNeil, M.A., et al. (2012) Comanagement of coral reef social-ecological systems, *Proceedings of the National Academy of Sciences*, 109(14): 5219–5222. https://doi.org/10.1073/pnas.1121215109

Claudet, J. (2018) Claudet's final word, *ICES Journal of Marine Science*, 75(3): 1177. https://doi.org/10.1093/icesjms/fsx140

Coad, L., Watson, J.E., Geldmann, J., Burgess, N.D., Leverington, F., Hockings, M., Knights, K. and Di Marco, M. (2019) Widespread shortfalls in protected area resourcing undermine efforts to conserve biodiversity, *Frontiers in Ecology and the Environment*, 17(5): 259–264. https://doi.org/10.1002/fee.2042

Cochrane, K.L., Rakotondrazafy, H., Aswani, S., et al. (2019) Tools to enrich vulnerability assessment and adaptation planning for coastal communities in data-poor regions: Application to a case study in Madagascar, *Frontiers in Marine Science*, 5: 505. https://doi.org/10.3389/fmars.2018.00505

Cole, D.H., Epstein, G. and McGinnis, M.D. (2019) Combining the IAD and SES frameworks, *International Journal of the Commons*, 13(1): 244–275. https://doi.org/10.18352/ijc.864

Coleman, E.A. (2009) Institutional factors affecting biophysical outcomes in forest management, *Journal of Policy Analysis and Management*, 28(1): 122–146. https://doi.org/10.1002/pam.20405

Coleman, E.A. and Fleischman, F.D. (2012) Comparing forest decentralization and local institutional change in Bolivia, Kenya, Mexico, and Uganda, *World Development*, 40(4): 836–849. https://doi.org/10.1016/j.worlddev.2011.09.008

Coleman, J.S. (2003) Norms as social capital, in Ostrom, E. and Ahn, T.K. (eds.) *Foundations of Social Capital*, Cheltenham, UK: Edward Elgar Publishing, pp. 133–155.

Costello, C., Gaines, S.D. and Lynham, J. (2008) Can catch shares prevent fisheries collapse? *Science*, 321(5896): 1678–1681. https://doi.org/10.1126/science.1159478

Cox, M., Arnold, G. and Villamayor Tomas, S. (2010) A review of design principles for community-based natural resource management, *Ecology and Society*, 15(4): 38. https://www.ecologyandsociety.org/vol15/iss4/art38/

Crawford, S.E.S. and Ostrom, E. (1995) A grammar of institutions, *The American Political Science Review*, 89(3): 582–600. https://doi.org/10.2307/2082975

Cumming, G., Campbell, L., Norwood, C., Ranger, S., Richardson, P., and Sanghera, A. (2022) Putting stakeholder engagement in its place: How situating public participation in community improves natural resource management outcomes, *GeoJournal*, 87(Suppl 2): 209–221. https://doi.org/10.1007/s10708-020-10367-1

Cumming, G.S., Epstein, G., Anderies, J.M., Apetrei, C.I., Baggio, J., Bodin, Ö.., Chawla, S., Clements, H.S., Cox, M., Egli, L., Gurney, G.G., Lubell, M., Magliocca, N., Morrison, T.H., Müller, B., Seppelt, R., Schlüter, M., Unnikrishnan, H., Villamayor-Tomas, S. and Weible, C.M. (2020) Advancing understanding of natural resource governance: A post-Ostrom research agenda, *Current Opinion in Environmental Sustainability*, 44: 26–34. https://doi.org/10.1016/j.cosust.2020.02.005

DeCaro, D.A., Janssen, M.A. and Lee, A. (2015) Synergistic effects of voting and enforcement on internalized motivation to cooperate in a resource dilemma, *Judgment and Decision Making*, 10(6): 511–537. https://doi.org/10.1017/S1930297500006975

DeCaro, D.A. and Stokes, M.K. (2013) Public participation and institutional fit: A social–psychological perspective, *Ecology and Society*, 18(4): 40. https://doi.org/10.5751/ES-05837-180440

Di Franco, A., Thiriet, P., Di Carlo, G. et al. (2016) Five key attributes can increase marine protected areas performance for small-scale fisheries management, *Scientific Reports*, 6: 38135. https://doi.org/10.1038/srep38135

Drazkiewicz, A., Challies, E. and Newig, J. (2015) Public participation and local environmental planning: Testing factors influencing decision quality and implementation in four case studies from Germany, *Land Use Policy*, 46: 211–222. https://doi.org/10.1016/j.landusepol.2015.02.010

Edgar, G.J., Stuart-Smith, R.D., Willis, T.J., et al. (2014) Global conservation outcomes depend on marine protected areas with five key features, *Nature*, 506: 216–220. https://doi.org/10.1038/nature13022

Epstein, G. (2017) Local rulemaking, enforcement and compliance in state-owned forest commons, *Ecological Economics*, 131: 312–321. https://doi.org/10.1016/j.ecolecon.2016.09.012

Epstein, G., Gurney, G., Chawla, S., Anderies, J.M., Baggio, J., Unnikrishnan, H., Villamayor Tomas, S. and Cumming, G.S. (2021) Drivers of compliance monitoring in forest commons, *Nature Sustainability*, 4: 450–456. https://doi.org/10.1038/s41893-020-00673-4.

Fidler, R.Y., Ahmadia, G.N., Amkieltiela, et al. (2022) Participation, not penalties: Community involvement and equitable governance contribute to more effective multiuse protected areas, *Science Advances*, 8(18): eabl8929. https://doi.org/10.1126/sciadv.abl8929

Frantz, C.K. and Siddiki, S. (2021) Institutional Grammar 2.0: A specification for encoding and analyzing institutional design, *Public Administration*, 99(2): 222–247. https://doi.org/10.1111/padm.12719

Frey, B.S. and Jegen, R. (2001) Motivation crowding theory, *Journal of Economic Surveys*, 15(5): 589–611. https://doi.org/10.1111/1467-6419.00150

Gibson, C.C., Williams, J.T. and Ostrom, E. (2005) Local enforcement and better forests, *World Development*, 33(2): 273–284. https://doi.org/10.1016/j.worlddev.2004.07.013

Gill, D.A., Mascia, M.B., Ahmadia, G.N., et al. (2017) Capacity shortfalls hinder the performance of marine protected areas globally, *Nature*, 543: 665–669. https://doi.org/10.1038/nature21708

Gutierrez, N.L., Hilborn, R. and Defeo, O. (2011) Leadership, social capital and incentives promote successful fisheries, *Nature*, 470: 386–389. https://doi.org/10.1038/nature09689

Harden-Davies, H., Amon, D.J., Vierros, M., et al. (2022) Capacity development in the Ocean Decade and beyond: Key questions about meanings, motivations, pathways, and measurements, *Earth System Governance*, 12: 100138. https://doi.org/10.1016/j.esg.2022.100138

Helmke, G., and Levitsky, S. (2004) Informal institutions and comparative politics: A research agenda, *Perspectives on Politics*, 2(4): 725–740. https://www.cambridge.org/core/journals/perspectives-on-politics/article/abs/informal-institutions-and-comparative-politics-a-research-agenda/11A6AA9DCAF579F5B72DA250728AA513#

Howe, C. and Milner-Gulland, E.J. (2012) Evaluating indices of conservation success: A comparative analysis of outcome- and output-based indices, *Animal Conservation*, 15(3): 217–226. https://doi.org/10.1111/j.1469-1795.2011.00516.x

Johns, C. (2021) Water governance indicators in theory and practice: Applying the OECD's water governance indicators in the North American Great Lakes region, *Water International*, 46(7–8): 976–999. https://doi.org/10.1080/02508060.2021.1996184

Knowlton, L.W. and Phillips, C.C. (2012) *The Logic Model Guidebook: Better Strategies for Great Results*, Thousand Oaks, CA: Sage.

Mahajan, S.L., Obiene, S., Ojwang, L., et al. (2024) Introducing Elinor for monitoring the governance and management of area-based conservation, *Conservation Biology*, 38(2): e14213. https://doi.org/10.1111/cobi.14213

Maynard-Moody, S. and Portillo, S. (2010) Street-level bureaucracy theory, in Durant, R.F. (ed.) *The Oxford Handbook of American Bureaucracy*, New York: Oxford University Press, pp. 252–277.

McGinnis, M.D. (2011a) An introduction to IAD and the language of the Ostrom workshop: A simple guide to a complex framework, *Policy Studies Journal*, 39: 169–183. https://doi.org/10.1111/j.1541-0072.2010.00401.x

McGinnis, M.D. (2011b) Networks of adjacent action situations in polycentric governance, *Policy Studies Journal*, 39: 51–78. https://doi.org/10.1111/j.1541-0072.2010.00396.x

McLaughlin, J.A. and Jordan, G.B. (2015) Using logic models, in Newcomer, K.E., Hatry, H.P. and Wholey J.S. (eds.) *Handbook of Practical Program Evaluation*, 4th edition, Somerset: Wiley, pp. 62–87.

Miller, D.C., Rana, P., Nakamura, K., Irwin, S., Cheng, S.H., Ahlroth, S. and Perge, E. (2021) A global review of the impact of forest property rights interventions on poverty, *Global Environmental Change*, 66: 102218. https://doi.org/10.1016/j.gloenvcha.2020.102218

Milman, A., Bunclark, L., Conway, D. and Adger, W.N. (2013) Assessment of institutional capacity to adapt to climate change in transboundary river basins, *Climatic Change*, 121: 755–770. https://doi.org/10.1007/s10584-013-0917-y

Moss, T. (2012) Spatial fit, from Panacea to practice: Implementing the EU water framework directive, *Ecology and Society*, 17(3): 2. https://doi.org/10.5751/ES-04821-170302

Nielsen, J.R. (2003) An analytical framework for studying: Compliance and legitimacy in fisheries management, *Marine Policy*, 27: 425–432. https://doi.org/10.1016/S0308-597X(03)00022-8

Obiero, K.O., Abila, R.O., Njiru, M.J., Raburu, P.O., Achieng, A.O., Kundu, R., Ogello, E.O., Munguti, J.M. and Lawrence, T. (2015) The challenges of management: Recent experiences in implementing fisheries co-management in Lake Victoria, Kenya, *Lakes & Reservoirs: Research & Management*, 20(3): 139–154. https://doi.org/10.1111/lre.12095

Ocampo-Diaz, N., Lopez, M.C., Axelrod, M. and Norris, P. (2022) Decentralizing the governance of inland fisheries in the Pacific region of Colombia, *International Journal of the Commons*, 16(1): 78–93. https://doi.org/10.5334/ijc.1131

OECD (2018) *Implementing the OECD Principles on Water Governance: Indicator Framework and Evolving Practices*, Paris: Organisation for Economic Cooperation and Development.

Orr, C.J., Adamowski, J.F., Medema, W. and Milot, N. (2016) A multi-level perspective on the legitimacy of collaborative water governance in Québec, *Canadian Water Resources Journal/Revue Canadienne des Ressources Hydriques*, 41(3): 353–371. https://doi.org/10.1080/07011784.2015.1110502

Ostrom, E. (1990) *Governing the Commons*, New York: Cambridge University Press.

Ostrom, E. (2005) *Understanding Institutional Diversity*, Princeton, NJ: Princeton University Press.

Ostrom, E. (2007) A diagnostic approach for going beyond panaceas, *Proceedings of the National Academy of Sciences*, 104(39): 15181–15187. https://doi.org/10.1073/pnas.0702288104

Ostrom, E. (2009) A general framework for analyzing sustainability of social-ecological systems, *Science*, 325(5939): 419–422. https://doi.org/10.1126/science.1172133

Partelow, S. (2018) A review of the social-ecological systems framework: Applications, methods, modifications, and challenges, *Ecology and Society*, 23(4): 36. https://doi.org/10.5751/ES-10594-230436

Persha, L., Agrawal, A. and Chhatre, A. (2011) Social and ecological synergy: Local rulemaking, forest livelihoods, and biodiversity conservation, *Science*, 331(6024): 1606–1608. https://doi.org/10.1126/science.1199343

Poteete, A.R., Janssen, M.A. and Ostrom, E. (2010) *Working Together: Collective Action, the Commons, and Multiple Methods in Practice*, Princeton, NJ: Princeton University Press.

Qiu, J., Wardropper, C.B., Rissman, A.R. and Turner, M.G. (2017) Spatial fit between water quality policies and hydrologic ecosystem services in an urbanizing agricultural landscape, *Landscape Ecology*, 32: 59–75. https://doi.org/10.1007/s10980-016-0428-0

Rahman, M.F. and Islam, K. (2021) Effectiveness of protected areas in reducing deforestation and forest fragmentation in Bangladesh, *Journal of Environmental Management*, 280: 111711. https://doi.org/10.1016/j.jenvman.2020.111711

Rajkumar, G., Narayan, G. and Mohammed Irshad, S. (2020) Transboundary disaster governance between India and Nepal-A perspective study on Kosi basin about risk governance dealing policies and communities, *Disaster Advances*, 13(4): 32–39.

Ramcilovic-Suominen, S. and Epstein, G. (2015) The impacts of deterrence, social norms and legitimacy on forest rule compliance in Ghana, *Forest Policy and Economics*, 55: 10–20. https://doi.org/10.1016/j.forpol.2015.03.006

Relano, V. and Pauly, D. (2023) The 'Paper Park Index': Evaluating marine protected area effectiveness through a global study of stakeholder perceptions, *Marine Policy*, 151: 105571. https://doi.org/10.1016/j.marpol.2023.105571

Ribot, J.C., Agrawal, A. and Larson, A.M. (2006) Recentralizing while decentralizing: How National Governments reappropriate forest resources, *World Development*, 34(11): 1864–1886. https://doi.org/10.1016/j.worlddev.2005.11.020

Rife, A.N., Erisman, B., Sanchez, A. and Aburto-Oropeza, O. (2013) When good intentions are not enough... Insights on networks of "paper park" marine protected areas, *Conservation Letters*, 6(3): 200–212. https://doi.org/10.1111/j.1755-263X.2012.00303.x

Ruano-Chamorro, C., Gurney, G.G. and Cinner, J.E. (2022) Advancing procedural justice in conservation, *Conservation Letters*, 15(3): e12861. https://doi.org/10.1111/conl.12861

Rustagi, D., Engel, S. and Kosfeld, M. (2010) Conditional cooperation and costly monitoring explain success in forest commons management, *Science*, 330(6006): 961–965. https://doi.org/10.1126/science.1193649

Ryan, R.M. and Deci, E.L. (2000) Self-determination theory and the facilitation of intrinsic motivation, social development, and well-being, *American Psychologist*, 55(1): 68–78. https://psycnet.apa.org/doi/10.1037/0003-066X.55.1.68

Schlager, E., Blomquist, W. and Tang, S.Y. (1994) Mobile flows, storage, and self-organized institutions for governing common-pool resources, *Land Economics*, 70(3): 294–317. https://doi.org/10.2307/3146531

Schlager, E. and Ostrom, E. (1992) Property-rights regimes and natural resources: A conceptual analysis, *Land Economics*, 68(3): 249–262. https://doi.org/10.2307/3146375

Scott, T. (2015) Does collaboration make any difference? Linking collaborative governance to environmental outcomes, *Journal of Policy Analysis and Management*, 34(3): 537–566. https://doi.org/10.1002/pam.21836

Sen, S. and Nielsen, J.R. (1996) Fisheries co-management: A comparative analysis, *Marine Policy*, 20(5): 405–418. https://doi.org/10.1016/0308-597X(96)00028-0

Siddiki, S., Heikkila, T., Weible, C.M., Pacheco-Vega, R., Carter, D., Curley, C., Deslatte, A. and Bennett, A. (2022) Institutional analysis with the institutional grammar, *Policy Studies Journal*, 50(2): 315–339. https://doi.org/10.1111/psj.12361

Stolton, S., Dudley, N., Belokurov, A., Deguignet, M., Burgess, N.D., Hockings, M., Leverington, F., MacKinnon, K. and Young, L. (2019) Lessons learned from 18 years of implementing the management effectiveness tracking tool (METT): A perspective from the METT developers and implementers, *Parks*, 25(2): 79–92. https://parksjournal.com/wp-content/uploads/2019/12/PARKS-25.2-Stolton-et-al-10.2305-IUCN.CH_.2019.PARKS-25-2SS.en_.pdf

Sze, J.S., Carrasco, L.R., Childs, D. and Edwards, D.P. (2022) Reduced deforestation and degradation in Indigenous Lands pan-tropically, *Nature Sustainability*, 5: 123–130. https://doi.org/10.1038/s41893-021-00815-2

Thiault, L., Jupiter, S.D., Johnson, J.E., Cinner, J.E., Jarvis, R.M., Heron, S.F., Maina, J.M., Marshall, N.A., Marshall, P.A. and Claudet, J. (2021) Harnessing the potential of vulnerability assessments for managing social-ecological systems, *Ecology and Society*, 26(2): 1. https://doi.org/10.5751/ES-12167-260201

Tilleard, S., and Ford, J. (2016) Adaptation readiness and adaptive capacity of transboundary river basins, *Climatic change*, 137: 575–591. https://doi.org/10.1007/s10584-016-1699-9

Trachtenberg, Z. and Focht, W. (2005) Legitimacy and watershed collaborations: The role of public participation, in Sabatier, P., Focht, W., Lubell, M., Trachtenberg, Z., Vedlitz, A. and Matlock, M. (eds.) *Swimming Upstream: Collaborative Approaches to Watershed Management*, Cambridge: MIT Press, pp. 149–210.

UNEP. (2021) *Progress on Integrated Water Resources Management. Tracking SDG 6 Series: Global Indicator 6.5.1 Updates and Acceleration Needs*, Nairobi: United Nations Environment Programme.

Viteri, C. and Chávez, C. (2007) Legitimacy, local participation, and compliance in the Galápagos Marine Reserve, *Ocean & Coastal Management*, 50(3–4): 253–274. https://doi.org/10.1016/j.ocecoaman.2006.05.002

5 Institutional complexity in linked socio-ecological systems – an approach to studying multi-level, cross-scalar contexts

Candace K. May

Introduction

Diversity is a critical component of flexibility and responsiveness in governing complex socio-ecological systems (CSESs); diversity is also critical to the ability to produce just outcomes (Calderon-Contreras and White, 2020). Individuals and groups of individuals who are engaged in and/or affected by governance processes have the potential to improve problem solving, programmatic correctives and responses to unexpected challenges. Diversity in race, class, gender, religion or other social categories, as well as knowledge systems, scale of commercial operations and user group representation (such as recreational, commercial or agricultural) and environmental interests, provides a range of material and non-material resources. This diversity also brings different skills, perspectives and approaches, and facilitates access to additional resources through social networks. All of these factors contribute to the adaptive capacity needed to respond to shocks and disturbances. However, opportunities and barriers to recognition, respect and inclusion in decision-making processes vary by stratified positionality[1] within and across institutional, organizational and everyday contexts. Scholars, researchers and practitioners must be able to examine opportunities, barriers and potentials for considering and including social diversity in discrete social, organizational and institutional components of governance systems to ensure just outcomes. The complex adaptive governance systems framework (CAGS-F) provides tools to examine how social structures and processes across levels and scales of CSESs differentially effect and are affected by diverse groups operating within and across institutional and organizational contexts.

The purpose of CAGS-F is to guide integrated socio-ecological work toward progressive social change for just and sustainable futures by revealing barriers, opportunities, processes and resources for considering and including diversity in experiences, knowledge and circumstances in governance of CSESs. CAGS-F was developed to provide a model of social complexity in CSESs for design and implementation of programs and policies, assessment or evaluation of unintended, unforeseen and/or undesirable outcomes, such as those that produce environmental injustices. The focus on governance in CAGS-F narrows attention

DOI: 10.4324/9781032624679-5

to the specific institutions, organizations and people involved in and/or affected by decision-making processes (May, 2022). CAGS-F pulls together Ostrom and colleagues' (Crawford and Ostrom, 1995; Schlager and Ostrom, 1992) institutional levels, Perrow's (1961, 1986) multi-level organizational goals, theories of practice (Bourdieu, 1990; Giddens, 1990) and critical realist, intersectional scholarship (Flatschart, 2017; Martinez, Martin and Marlow, 2014). Bhaskar's (2008) critical realism is used to represent the latent world of invisible, informal institutions to complement Ostrom's approach, which priviledges formal institutions. Gunderson and Holling's (2002) panarchy heuristic serves as the foundational logic for understanding and emphasizing the dynamic hierarchically ordered, cross-scalar, multi-level connections, pressures and constraints to change across the institutions, organizations and people of governance systems.

Adaptive cycles and panarchy

Gunderson and Holling (2002) developed the panarchy model to explain resilience – the ability of a system to resist, adapt or transform to a new resilience regime in response to internal (i.e., political turmoil) or external (i.e., natural disasters) shocks and disturbances. The outcome of pressures and constraints generated by interdependent but discontinuous adaptive cycles operating at multiple levels and scales produces resilience, adaptation or transformation (May, 2022). Figure 5.1 illustrates Gunderson and Holling's (2002) adaptive cycle and panarchy model with altered, social scientific terminology (May, 2021a). Crisis (Ω) reflects a shock or disturbance from natural disasters, economic or political crises, demographic change, technological innovations or

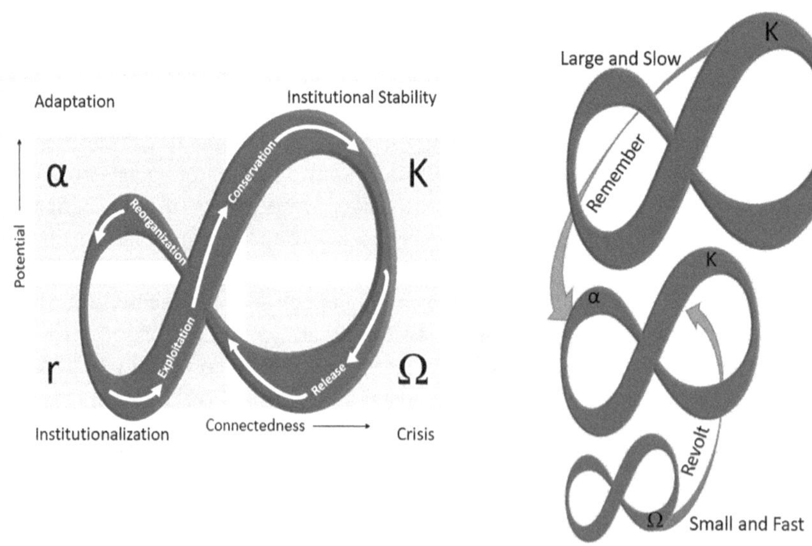

Figure 5.1 Model of adaptive cycle and panarchy heuristic. Adapted from Gunderson and Holling (2002).

development projects. Adaptation (α) refers to actual or potential responses to crises, such as disaster preparedness planning, social mobilization or environmental conflict resolution. Institutionalization (r) is a process where adaptations (α) become patterned and recurring in infrastructural development processes, policies or final plans. Institutionalized stability (K) is created when understandings, expectations and actions over time become routinized.

Panarchical systems consist of multiple levels and scales of adaptive cycles (Figure 5.1), which vary in response to pressures for change. The cycles at higher levels change at a slower pace and exert greater force in constraint and authority at broader spatial, temporal, jurisdictional or conceptual/analytical scales. Cycles at lower levels change faster and exert less constraint and authority across smaller scales. The 'remember' and 'revolt' connections in Figure 5.1 depict top-down constraints and bottom-up pressures for change. The larger and slower components of the hierarchy provide stability and encourage a return to a stability state after a disturbance. However, adaptive responses at lower levels and scales have the potential to cause or resist and divert change from higher levels and scales (Gunderson and Holling, 2002).

CAGS-F theorizes that institutions, organizations and human practices as cross-scalar, interconnected panarchical systems, where discontinuous adaptive responses within and across the complex system precede occur conterminously and follow socio-ecological shocks (May, 2022). It is important to note that orgainzations are treated as separate and distinct from institutions in CAGS-F. Each component is defined in Figure 5.2 and explained below with examples

Institutions		Organizations	Conceptualizations and Examples
Informal	Formal		
Real			Nearly unobservable, but discernable from how they shape what activities, processes, and tendencies occur or are possible across the system (e.g., Racism, Sexism, Misogynoir; Deep Seated Values of Individualism and Private Property, Communalism, Relationality; Development Trajectories, i.e., Modern vs. Traditional, Knowledge Systems.)
	Constitutional		Designate authority relations at the highest formal institutional level, with influence across lower levels and scales (e.g., laws, legislation, enabling statutes.)
Actual			Events, relations and factors require a degree of concerted effort to discern and explain (e.g., Differential life chances and livelihood consequences from institutional and organizational barriers and opportunities for recognition, respect, and inclusion, including differential valuations of social, cultural, and economic capital resources.)
	Collective Choice		Processes and procedures for decision-making and the roles of actors involved (e.g. Policy Protocols and Processes; Court and Legislative Studies Protocols, Proceedings, Outcomes.)
		Official	The general purpose of an organizations as set out in charters, mission statements, annual reports, public statements by key executives, and other authoritative pronouncements.
		Operative	Actual operating policies of the organization which delineate means and resources and prioritize objectives in achieving official goals.
		Unofficial	Operational and informal organizational aspects, which evolve with the preferences, interpretations, capabilities, and power struggles of personnel in everyday problem solving and role fulfillment.
Empirical			Events, relations, and factors are more easily observed and measured (e.g. inter-personal racism; efforts to control access/use/allocation, i.e. erection of blockades; organizing of public forums for information sharing, recruitment, and mobilization; absence from venues and consideration.)
	Operational		Guiding rules for every day, practical activities, decision-making, and problem-solving processes for individuals or small collectivities (e.g. Innovative rulemaking in access and use of resources)

Figure 5.2 Conceptualizations and examples of panarchical informal and formal institutions and organizations.

in separate sections. A full example application of CAGS-F is provided through a case study of North Carolina Blue Crab Governance.

Institutional panarchy

CAGS-F incorporates Ostrom's (2005) three primary institutions, the constitutional, collective choice and operational, as a panarchy of formal institutions (May, 2021a).[2] Constitutional rules determine the structure and processes of collective-choice and operational rules, including amendments and who can participate. Collective-choice rules determine processes and procedures for decision-making and the roles of actors involved, which in turn determine operational rules. Operational rules govern everyday practices, decision-making and problem-solving processes, and monitoring and sanctioning occur through organizational disciplinary tactics. In contrast to Ostrom (2005) and colleague's (Crawford and Ostrom, 1995; Schlager and Ostrom, 1992) initial theory, the institutional panarchy allows for bottom-up change and encourages attention to divergence between the institutional levels to detect innovation and tension that may indicate pressures or resistance to change. For example, May (2021a) showed how people created innovative ways to access newly expanded public waters overlying private agricultural lands at the operational level in South Dakota, USA (see Figure 5.3). Landowner attempts to control access to their lands resulted in adaptative court decisions and House Bills at the collective choice level. Collective choice decsions resulted in new and altered statutory laws at the constitutional level. In this case, emergent responses to expanded public waters overlying private lands at the level of everyday activities filtered up to effect changes in statury laws.

Formal	Adaptive Cycle			
Institutions	Crisis (Ω)	Adaptation (α)	Institutionalization (r)	Institutionalized Stability (K)
Constitutional	Buying/Leasing Public Trust Resources Indeterminant Recreational Rights of Use of Waters Changing Hydrology Intersectionality & Equity	TBD	TBD	Changes in State Laws
Collective Choice	Landowner lawsuits against GFP	GFP Closed Access Points to waters. Legislative Summer Study	HB 1001	Resumed Access to Waters New GFP responsibilities to negotiate with landowners
Operational	1990s Wet Cycle Inundated Farmland Wetland Expansion	Farmers expanded, were displaced, or diversified. Increased public use of expanded waters	GFP Developed Infrastructure for Public Access to Waters	Infrastructure development normalized public use expectations

Figure 5.3 Institutional Panarchy in South Dakota.

Organizational panarchy

The institutional outcomes noted above also involved multi-dimensional organizational processes. An organizational panarchy is conceptualized using Perrow's (1986) three levels of goals (official, operative and unofficial) internal to organizations, which change in response to external natural, social and institutional environments and as a result of the internal discretion of people filling roles. Delineation of organizations as separate from institutions in CAGS-F is important. Organizations are actors, as well as arenas for action. Institutions are rules. Institutions are neither actors nor the arenas where actions occur. Organizations enter collective choice arrangements in governance systems, often with a higher degree of authority than local individuals or groups and lower degree of authority than legislation and laws. Organizations are also arenas for collective action processes. Organizations operate across institutional levels, at operational levels within communities and at constitutional levels to change laws and policies. However, organizations are not unified, homogenous entities. The more-or-less structured roles and hierarchical levels of authority characteristic of organizations do not fully control for diversity in human interests, skills, ingenuity, cooperation and conflict in fulfilling missions, goals and objectives (Perrow, 1986; Selznick, 1949).

An example is mapped out in Figure 5.4. Public innovations in accessing public waters overlying private lands in South Dakota at the operational institutional level spurred a change in the state agency responsible for managing public natural resources, South Dakota Game, Fish and Parks (GFP) (May, 2021a, 2022). At the unofficial organizational level, GFP staff facilitated public access the waters outside statutory roles and responsibilities, but also in an absence of regulatory restrictions on their abilities to do so. These actions resulted in landowner use of the court system and concomitant collective choice rules and changes to the organizational operative goals of GFP. Negotiated rule-making processes between the state agency and landowners were added to agency roles along with responsibilities to facilitate public use of waters on private land. Ultimately these changes resulted in higher-order

	Adaptive Cycle			
Organization	Crisis (Ω)	Adaptation (α)	Institutionalization (r)	Institutionalized Stability (K)
Official				Management of Public Access/Use of Waters
Operative	Landowner Lawsuits Summer Study	Closed Access Points Stopped Encouragement/Management Actions		
Unofficial	1990s Wet Cycle Inundated Farmland Wetland Expansion	GFP Facilitated Public Access to Waters – Stocked Fish, Built Docks, Added Parking	Infrastructure Maintenance Public Use Encouragement and Management	

Figure 5.4 Organizational panarchy in South Dakota.

institutional changes at the constitutional level and official organizational change. The end results were legal changes in how public waters overlying private lands were defined, town responsibilities in managing road access and state agency responsibilities for managing public access to these water.

Informal institutions and multiple panarchical realities

The discussion so far has focused on formal institutional and organizational structures and processes, which are manifest and easily observable in documented legislation, laws, policies, organizational mission statements, meeting minutes, workload documents, contracts and other documents. In contrast, informal institutions are latent, undocumented, implicit, often tacit norms, understandings and expectations (North, 1990), which infuse the logic of laws, policies and organizational forms and functions, as well as the disciplinary tactics of everyday social relations. Informal institutions are conceptualized through Bhaskar's (2008) stratified realities system, which is highly compatible with Gunderson and Holling's (2002) panarchy heuristic. Bhaskar's (2008) premise is that the social world is constituted by hierarchically related social realities that operate continuously, regardless of immediate effect and irrespective of any one person's acknowledgement. We know the underlying causal mechanisms of informal institutions are real and we can study them from their effects on human behaviours, circumstances and life chances that are not readily or entirely explained by manifest, formal institutional and organizational features.

Bhaskar (2008) conceptualized three levels of reality, which are labelled down the left-hand side of Figure 5.2. The levels range from empirical to latent in ease of observability. At the top of the panarchy are *real*, nearly unobservable tendencies, which provide weight and meaningful structure to the social world. The mid-level consists of *actual* events, which require a somewhat concerted effort to discern and explain; and empirical events at the lowest level are easily observed and measured. The panarchy heuristic adds dynamic processes of change and resistance across each level and scale of reality. The real level changes at a slower pace and exerts greater force in constraint and authority at broader scales. Cycles at the empirical level are faster and exert less constraint and authority across smaller scales. However, realities are not just stratified, but multiple.

Following Bhaskar (2008), critical realist, intersectional scholars (Flatschart, 2017; Martinez, Martin and Marlow, 2014) discuss abstract categories, such as racism, sexism and classism, as well as intersectional mechanisms (such as the notion of misogynoir, the misogyny directed specifically at black women and girls) as *real*, emergent structures. These structures stratify individuals and groups within social hierarchies with *actual* consequences for life chances, which manifest as *empirical* tedencies in life conditions and everyday interactions, conflicts, challenges and successes (Martinez, Martin and Marlow, 2014). For example, while U.S. constitutional institutions and formal organizational goals bar race and gender as a basis for discrimination, racism and sexism persist across different levels of institutions, organizations and relationships. However, these social facts manifest as disparities in public health, income, occupations and opportunities for participation in decision-making processes; and, they are evident through everyday interactions, relationships, challenges, successes and collective mobilization for change toward racial and gender equity.

Critical intersectionality reveals how informal institutional processes produce differential outcomes across groups regardless of the stated procedures and imperatives of laws and organizational missions. Real relations of power, privilege and domination exist outside the activities or recognition of any one individual actor and are produced and reproduced through organizations and institutions (Ray, 2019). For example, Taylor (2015) demonstrated how environmental injustices were partly the result of a lack of diversity among those holding positions of authority in environmental nongovernmental organizations and foundations. The composition of those in authority matters. For example, Rothstein (2017) showed historical organizational goals of discriminatory housing practices, such as redlining, which is now illegal, created persistent residential segregation and inequities in life chances that persist today. The legacy of these institutional structures is reproduced or challenged by people carrying out unofficial organizational goals. Intentionally or unintentionally, people reinforce or change informal institutional structures and processes that perpetuate injustice through how their preferences, knowledge and experiences affect organizational performance and broader responses to pressures for change.

People and panarchical practices

The formal domains of institutional and orgainzational logics are manifest, easier to observe with the assistance of traditional theories, frameworks and methods. In contrast, latent institutional, organizational and differential relational structures, processes and practices require the perspectives and experiences of groups from diverse backgrounds and experiences (May, 2022). Bourdieu's (1990) practice theory is helpful in this regard. The focus is on the emergent relations and practices of diverse and stratified social groups with differential capacities navigating the institutional and organizational dynamics of CAGSs. Therefore, data collection and analyses of social groups should proceed from the individual level to social classes[3] representing groups with similar life chances as per the advice of Bourdieu (1985). The incopration of this approach with a constant comparative assessment of manifest institutional and orgainzational structures and processes illuminates the invisible aspects of institutional and organizational panarchies and facilitates an examination of adaptive cycles across everyday practices, roles, responsibilities and differential access to and use of resources (empirical and operational institutional fields and unofficial organizational goals); relations within collective action contexts to other groups and policy structures and outcomes (the actual and collective choice institutional fields and operative organizational goals) and deeper

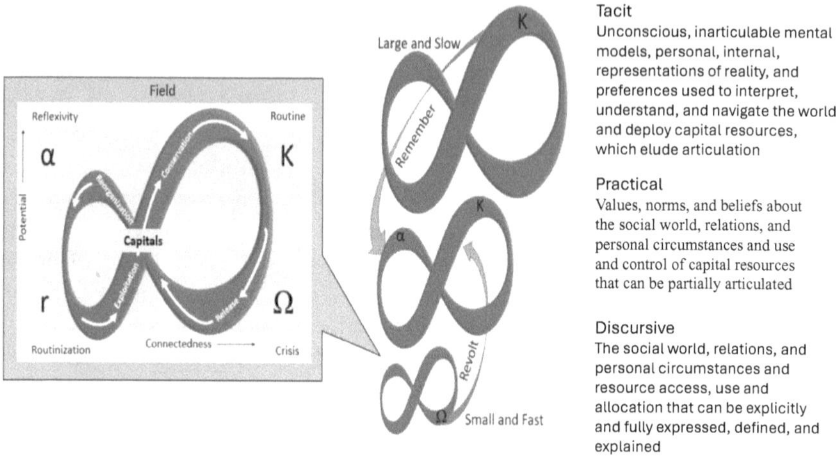

Figure 5.5 Conceptualization of people, capacities and change.

mechanisms of power and privilege (real and constitutional institutional fields and official organizational goals).

Figure 5.5 depicts a conceptual model of panarchical human practice that incorporates theories of practice from Bourdieu (1990) and Giddens (1990). Bourdieu's (1990) theory posits that the activities of individuals or social classes are enabled and structured by their habitus and the resources they control and/or have access to across fields. Resources include economic capital (equipment, property, credit or money), cultural capital (skills, knowledge, ways of seeing, understanding and doing) and social capital (the resources accessible through social networks, including status and respect as well as other people's economic, social and cultural capital). People use and deploy capital resources according to dispositions, perceptions and appreciations (habitus) in competitive contexts (fields) defined by stratified positions of prestige and explicit and implicit rules, boundaries and objectives for the use and valuation of capitals. For example, small-scale fisherpeople's habitus and capital resources emerge from and are highly useful for their livelihoods as well as designing and enforcing informal institutional understandings for use, allocation and access to fishery resources amid their community contexts (May, 2012). However, fisher capital resources are often undervalued and less effective in formal fishery policy processes relative to competing user groups due to power imbalances, a legacy of formal institutional regulation of commercial fishing, and informal institutional undervaluation of the environmental impact of recreational fishing activities (May, 2016).

People, along with their capital resources and habitus, change through interactions and experiences, potentially leading to widespread consequences for institutional and organizational domains. Figure 5.5 illustrates an adaptive cycle for habitus, where new, challenging or altered experiences or social relations spur reflexivity, routinization of an adaptive response and a potential consequent change in habit or routine. Reflexivity creates opportunities for change in the same way crises spur adaptive cycles (Gunderson and Holling, 2002) or threatened ontological security has the potential to disrupt routines in social and material environments (Giddens, 1990). Bourdieu's (1990) theory emphasizes how practice and reflexivity has the potential to change consciousness, relations and resource access in ways that can reinforce, challenge or transform resilient latent and manifest assumptions, preferences and understandings embedded in institutional, organizational and human logics.

Practice and reflexivity are panarchical. Tacit assumptions, preferences and understandings signify the existence and workings of broad scale informal institutions and are embodied to different degrees in habitus (peoples' dispositions, perceptions and appreciations) (May, 2022). These structures exist before contemporary contexts and are reflected to varying degrees in formal institutional and organizational logics. However, they can evolve through practice and reflexivity among individuals and social groups. At the broadest scale of the panarchical model is tacit knowledge – the unarticulated knowledge, constituted by the intuition, preferences, beliefs and values that people form because of experiences and socialization, which serves as the filter for interpretations, understandings and the basis for most individual decisions and practical activity. Between tacit (unarticulated) and discursive (articulated) knowledge lies everyday practical consciousness (Giddens, 1984). At the highest level, tacit knowledge is unconscious and inarticulable, while practical consciousness at the mid-level is semi-articulable and discursive consciousness at the lower level with narrowest scope reflects what people can fully articulate regarding the world and their actions and circumstances (May, 2022). What people understand and can articulate has the potential to change through the iterative processes of experimentation and learning in defining problems, identifying solutions, implementing strategies and assessing outcomes in group settings (May, 2022). In summary, CAGS-F emphasizes the relational processes of reflexive panarchical practice (the stratified positioning of actors based on power and privilege – resources and valuations) in and across fields (formal institutions and organizations) for how latent institutions and tacit and explicit assumptions, preferences and understandings are and can be reinforced, challenged or transformed as discursive consciousness expands.

Change from the bottom-up or resistance from the top-down

Theoretically, change in relationships and experiences cause a change in people's economic capital (money, credit, tools and technology), social capital (resources accessible through social relations) and cultural capital (skills, knowledge, ways of knowing) (May, 2022). The degree of trust, reciprocity and mutual understanding and respect underlying relationships and relationship building cause increases, decreases or qualitative changes to the resources people have or can access. For example, a comparative study of coastal community resilience in the United States showed how recognition and respect were crucial factors in determining whether historical small-scale commercial fishing communities experienced resilience or transformation. In Louisiana, diverse, stratified groups cooperatively pooled differential capital assets to respond and recover from socio-ecological hazards in ways that built on pride in the local history as a commercial fishing community, while in North Carolina, transformation from commercial fishing to recreational and tourist communities occurred through conflict, distrust and exclusion of small-scale fishers (May, 2021b). In these cases, recognition and respect, or the lack thereof, were more than individual attributes, they were embedded in informal and formal institutional and organizational logics (May, 2021b).

Collaborative learning and co-productive governance processes (Abidi-Habib and Lawrence, 2007; de Kraker, 2017; Fernández-Giménez et al., 2019; Tschakert et al., 2014; Turnhout et al., 2020; Wyborn, 2015) have the potential to create reflexive practice and bottom-up change, but they can also reproduce inequities. In North Carolina fisheries, small-scale commercial fisherpeople have ample opportunity to participate in collaborative governance processes, but most fisherpeople have historically chosen to practice active non-participation because of perceived and actual limitations in political efficacy (May, 2012). Non-participation, whether by choice or exclusion, indicates broader structures and processes at play. In the North Carolina case, these issues were evident in the differential valuation of recreational versus commercial fishing activities in regulatory policies and legislation; the limitations of the budgets, roles and responsibilities of fisheries agency staff; and empirical, actual and real preferences and imperatives for amenity-led development trajectories, which caused displacement of commercial fisher communities, infrastructure and livelihoods.

Figure 5.6 continues the South Dakota case for informal institutions. Indigenous people to the region[4] and representatives from Hutterite Colonies[5] were absent from processes, deliberations and consideration (media, public forums, lawsuits and the Legislative Summer Study) to determine access and use of public water overlying private lands (May, 2021a). Farmers and recreational interest groups were well represented and benefited from further capital resource development whether they directly participated in mobilizing support, leveraging social connections (including among government officials and state legislators) or engaged in public political forums or not. In addition, broader imperative at the *real* level privileged the sanctity of private property and political economic imperatives, which created *actual* pressure for a speedy resolution. The underlying, *empirical* logic of most discussions centred on the economic importance of agriculture and nature-based activities for the state economy. These imperatives were further reinforced by the constitutional adaptations designed to manage a public trust resource, while maintaining the integrity of individualism and private property rights.

Informal Institutions		Adaptive Cycles		
	Crisis (Ω)	Adaptation (α)	Institutionalization (r)	Stability (K)
Real	Colonization Racism & Genocide Immigration for Religious Freedom	Settlements Reservations Colonies	Built Environment Segregation	Racism & Prejudice Euro-American Individualism and Private Property Native American Relationality and Responsibility Hutterite Communalism
Actual	**Farmers** Livelihood consequences of inundation and public use of land.	Control use and access to land through Courts and Local Governments	Protection of Property Rights/Rights to Exclude Public use	Reinforced deep seated values of Individualism and Private Property
	Recreational Interest Groups Economic and human welfare consequences of expanded recreational opportunities.	Participation in determining use and access to waters	Mobilization of Support, Recognitions of Issue in Law	Reinforced mission to promote/increase public access to nature
	Indigenous Populations No Information	No Information	No Information	No Information
	Hutterite Colonies No Information	No Information	No Information	No Information
Empirical	**Farmers** - Flooded Agricultural Lands	Efforts to control access/use, i.e., erection of blockades	Maintenance, Changed, and New Livelihoods	Greater Representation in Laws
	Recreational Interest Groups - Increased Recreational Opportunities	Organized public forums for information sharing, recruitment, and mobilization of public support	Strengthened relationships and support for public access to recreation opportunities	Greater Representation in Laws
	Indigenous Populations No Information	No Information	No Information	No Information
	Hutterite Colonies Little Information: 25% of farmland of one colony lost	No Information	No Information	No Information

Figure 5.6 Informal institutions – panarchical reality systems in South Dakota.

The simple inclusion or omission from consideration of diverse groups in political deliberations, media reports and data indicates how generalized, tacit institutional norms, understandings and expectations operate through formal institutional and organizational domains to define who, how and what matters, as well as the potential for reproduction, challenge or transformation of the system.

An example application: governance of North Carolina Blue Crab

The governance of Blue Crab in North Carolina supplies another example of bottom-up and top-down panarchical pressures. The blue crab fishery is the largest and most valuable commericial fishery in North Carolina, and is the largest blue crab fishery in the United States (NCDMF, 2020). In 2009, commercial fishers in a small North Carolina community described the blue crab fishery as in a state of crisis from overharvesting and lack of regulations (May, 2011). They believed there should be limits on the number of crab pots each fisher could use, closed seasons and, most importantly, restrictions on harvesting egg-bearing females. As one fisherwoman explained,

> everybody around here – if we catch one, we let them go. If she is blossoming and that apron is burst on and full of eggs we will not throw them in the culls. I say hey that crab ain't worth but $0.10 anyway. I let her go – that is 10 million babies. That's our future.

A couple of fishers recounted how they would release the egg-bearing crabs they had harvested in other areas into the creeks adjacent to their community. According to these fishers, these actions protected the female crabs from being sold to the crab houses and increased the populations of crab in their community for the next years harvest. However, fisheries management personnel were unresponsive to crabber concerns, perspectives and experiences. See Table 5.1 for an application of CAGS-F to this case.

In 2009, administrators and fishery scientists said they knew of the fisher concerns and actions regarding blue crabs. However, they did not see a need for further regulations. They were ambivalent about the effects of moving the egg-bearing crabs from one area to another on population dynamics. At the very least, adminstrators and scientists did not see these activities as harmful; and, at worst, they thought it was economically wasteful because of the income fisher were foregoing. Justificatiaons for a lack of regulatory action included best available science and management responsibilities as defined by existing fishery management plans (FMPs in 1998 and 2004).[6] Resources for management actions, time being one of the most important, were also a consideration. FMPs delegate what needs to be done for sustainable management of fish populations and, as a consequence, distribute resources – time, funding, human capital, etc. – for management actions accordingly.

Fisher concern and actions did not produce enough pressure on the organizational level to spur change in organizational actions (regulations or research) in 2009. Amendments to the Blue Crab FMPs were not implemented until 2013 and 2016 (NCDMF, 2020). In 2013, fisheries management recogntion of diminshing crab harvests[7] spurred regulatory changes, which replaced set size limits for mature female crab with an adaptive management framework involving an annual assessment. In 2015, the adaptive management framework indicated a need for regulatory action, resulting in regulations implemented in 2016 that limited harvesting of several categories of female crabs (NCDMF, 2020). In 2018, the stock assessment determined that blue crabs were overfished and that overfishing was occurring.

Hence, in the absence of constitutional rules and a lack of collective choice actions, and in the context of a lack of attention at the organizational level, fishers devised their own operational rules for blue crab. In 2009, fisheries administrators and scientists did not have an interest in acting at the unofficial organizational level. They were also restricted from taking action by constraints from operative goals (set timelines for FMPs, distribution of roles and

Table 5.1 CAGS-F for North Carolina blue crabs

Institutions		*Organizations*	*Adaptive cycles*			
Informal	*Formal*		*Crisis (Ω)*	*Adaptation (α)*	*Institutionalization (r)*	*Institutionalized stability (K)*
Real			Socio-ecological & cultural decline			Fisher–value of livelihoods Management prioritization of science over local knowledge
	Constitutional		1996, federal response to overfishing			1997, State Legislation – protect ecological and cultural functions of fisheries through use of best available science
Actual			Overfishing Crab population decline	Crabber desire for regulations		Devaluation of fisher knowledge
	Collective choice		Commercial fisher active non-participation	Greater involvement of recreational, environmental, and large-scale commercial groups	Strengthening of recreational, environmental, and large-scale commercial fisher groups	Extensive collaborative fisheries governance framework
		Official	2018, official designation of overfished and overfishing			NCDMF mission – sustainable management of fisheries
		Operative	2013, statistical proof of crab population declines	Regulations and adaptive management	2016 regulations	Five year FMPs
		Unofficial	No recognized need for action			Restricted actions re limited resources and investment in FMP process
Empirical			Reduced crabber incomes	Crabber discard or relocation of crabs		
	Operational		2009, reduced harvests	Harvesting rules	Self-governance among crabbers	

responsibilities, and resources, i.e. time and funding of human capital). Here, the operative goals created a gap between the actions and concerns of fishers and the official goals of the NCDMF to end overfishing, as defined by state legislation (at the constitutional institutional).

Reasons for a lack of action in response to the early activities and concerns of commercial fishers beyond organizational constraints at the operative and unofficial goal levels involved a divergence in valuation of science versus local ecological knowledge (May, 2016) and the time scale differences in observations and actions between local communities and bureaucratic management agencies. Administrators and scientists were confident in the fisheries science at hand and in the organizational protocols in place, which operated at five-year intervals for FMP development. Preferences that privilege science over local knowledge (i.e., cultural capital) and formal means of management over everyday operational activities were legislated at the constituitional level (May, 2016). However, these preferences also have their roots in the informal institutional level of the real. These preferences are often unacknowledged, tacit representations with a broad scope of influence resulting in actual consequences (overfishing), empirical actions (concern and actions of local fishers) and subsequent adaptations in operational rules (fisher imposed limitations on harvesting and innovative release of egg bearing crabs). At lower temporal and closer proximity scales, the ability of local fishers to adapt operational rules to empirical changes in socio-ecological contexts is faster (not necessarily more accurate) than organizational entities at the collective choice level, which often involves a redistribution of resources, personnel roles and responsibilities, and scientific protocols and potentially changes in constitutional level allocations. This example shows how CAGS-F can be used to locate where human will for change meets resistance in order to explain socio-ecological outcomes and plan for improved responses in the future.

Using CAGS-F – a starting point

The structure of CAGS-F employs analytic dualism (Archer, 1995) to provide a way to identify where opportunities, barriers and leverage points for change exist, at what level and scale, and at which point in discrete adaptive cycles across institutional and organizational domains. Analytic dualism involves conceptual separation of the coupled human, non-human and natural components of socio-ecological systems. Analytic separation of institutions, organizations and people as distinct components of governance systems allows us to examine how each operates at different scales and levels. This approach helps detect goal displacement and drift, and identify how mechanisms of power and privilege operate through formal and informal institutions. Such analysis informs how organizations respond to social and ecological circumstances and how they implement and enforce their goals and objectives. Separating out legislation, policy processes and organizational operations across scales and levels, including the actions of people fulfilling official roles, allows for identification of variation in barriers, opportunities, leverage points and resources, including allies, for change. Analytic dualism increases the utility of CAGS-F as a conceptual tool for identifying where, at what level and scale, who is included and/or adversely affected, and at which point in the adaptive cycle opportunities, barriers and leverage points exist to optimize design, planning, programming and implementation of initiatives, or evaluate unintended and unforeseen or less than successful outcomes (May, 2022).

Pressures, resistance, and the need for institutional and organizational change are revealed in the constraints and opportunities, processes and outcomes of governance systems for groups who engage in, as well as those who are affected but absent from governance processes. After identification of the issue, the first step is to identify important groups by asking: what is the decision-making context/s; who is affected; who is involved and considered and who is affected but not involved or considered in decision-making processes that are creating/managing/seeking to resolve the issue at hand? Once initial individuals and groups have been identified, a series of broad questions to assess the system follow.

1 Questions into the formal institutional domain include:

   - How do groups reinforce, challenge or work to transform operational rules to control access, use and allocation of resources?
   - What are the barriers and opportunities at the collective choice level for participating in decision-making?
   - How do operational and collective choice level rules align with or contradict the constitutional level?

2 Questions into the organizational domain include:

   - How do people fulfilling roles within organizations fulfil responsibilities?
   - Which resource user-groups are considered in decision-making, engagement or regulation processes, with what resources, and under what directives?
   - How do the activities of people fulfilling organizational roles align with or contradict operative and official goals?

3 Questions to guide an examination of informal institutions and the practices of people operating from multiple stratified realities include:

   - What capital resources do groups have?
   - How are various capital resources employed, why and with what effectiveness in different fields relative to other groups?
   - What are the barriers and opportunities for recognition, respect and inclusion of different groups, including how different capital resources are valued within the system?

These questions are not exhaustive, and additional groups may be identified through the process of data collection and analysis. The questions above provide a starting point for collecting the data needed for comparative analyses of groups across organizational and institutional domains. The process of analysis will reveal connections representing contradictions, similarities and differences, and causal relationships in differential opportunities, constraints and resources that affect the recognition, representation or participation of individuals and groups across institutional and organizational systems.

A modified multilevel intersectional approach (Winker and Degele, 2011) is a useful analytic tool to examine institutional and organizational logics from the perspectives of individuals and groups with multiple, differential realities. The following eight steps are based on a case study approach using field observations, interviews and document review, but can be adapted depending on context, issues and social groups and, potentially, iterative processes of data collection and analysis depending on research design.

1 Identify all categories of differentiation and socio-environmental relations for individuals that serve as implicit or explicit 'self-positioners' in everyday practices, roles, responsibilities and access, use and allocation of resources.
2 Identify dominant/hegemonic, socio-political or culturally specific norms, values and ideologies related to categories of differentiation and socio-environmental relationships exhibited in experiences of different people with access, use and allocation of resources.
3 Link social structures to social practices through actions and/or references that agree with, oppose or challenge, or demonstrate variable degrees of inclusiveness or indifference to institutions, organizations or processes.
4 Conduct an iterative analysis of compatibilities and contradictions across experiences, knowledge and understanding, and access to and use of resources.
5 Cluster individuals into groups based on internal homogeneity and external heterogeneity of social practices identified in Step 4, leaving considerations of race, class or gender or dominant ideologies until after initial clusters based on practices, life chances and resources have been constructed.
6 Collect further data to examine structural features and group representation in laws, policies, bureaucratic organizations or political processes.
7 Incorporate data from media, advertising, photographs or written texts to understand norms, values, ideologies and relevant contexts to a greater extent.
8 Elaborate interrelations among dimensions of diversity, inclusion and equity in practices across socio-ecological contexts and fields.

The above questions and analytic steps provide a start for creating a synthesis of connections across institutional and organizational domains to produce a stratified, panarchical system for each individual case. The individual cases can be aggregated to create social groups by similarities in life chances. Comparative analyses across the panarchical systems for each group reveal connections representing contradictions, similarities and differences, and causal relationships in resistance and change across the CAGS.

Conclusion

CAGS-F provides basic guidance, organization and conceptualization of concepts and terms for diagnostic, descriptive and prescriptive inquiry into how institutional and organizational structures and processes and the activities of diverse groups of people across levels and scales produce differential outcomes, pressures, constraints and leverage points for achieving greater socio-ecological justice and sustainability. CAGS-F was built from a synthesis of scholarship on institutions, organizations and intersectionality and stratification with the panarchy heuristic within a critical realist orientation. The focus on governance narrows attention to the specific institutions, organizations and people involved in and affected by issue-specific contexts. The framework employs analytic dualism to help scientists, scholars and practitioners examine discrete multidimensional pressures, constraints and feedback effects across the institutions, organizations and people of governance systems. Intersectional and stratification theories allow us to understand and increase inclusion of diverse realities and differential capacities. Critical realism helps to make the invisible, unacknowledged and disregarded forces and relations of power and privilege that produce unintended and undesirable outcomes and injustices, visible to contribute to progressive goals.

Diversity is a critical component of flexibility and responsiveness in governing CSESs, as well as the ability to produce just outcomes (Calderon-Contreras or White, 2020). Diverse individuals and groups, whether by resource user-group status, commercial scale, knowledge system or race, class, gender or other social category, provide diverse resources. Problem solving, program design and adaptive responses to unexpected challenges are improved by diverse skills, ways of seeing, being, understanding and doing, and means of accessing additional resources through social networks (Calderon-Contreras and White, 2020). However, opportunities and barriers, including dispositions, perceived and actual external valuation of knowledge, skills, relations and other resources vary by stratified positionality within and across institutional, organizational and everyday contexts. Working for just outcomes requires the ability to adequately assess opportunities, barriers and potentials for considering and including social diversity in/and across social, organizational and institutional components of CSESs system. CAGS-F provides tools to examine how diverse groups operating within and across institutional and organizational contexts differentially affect and are affected by social structures and processes across levels and scales of CSESs (May, 2022). An understanding of equity issues is important for being able to 'see' injustices on the ground, but we must also be able to recognize people's resourceful capacities, experiences and perceptions outside our expectations. Following Bourdieu (1990), we can build from empirical social practices to a comparative examination of institutional and organizational processes and outcomes to discern deeper dimensions that structure opportunities and potentials for socio-ecological justice. In this way, CAGS-F provides guidance, and an orientation sensitive to diversity, power and privilege without closing the research process to surprising and unexpected results.

Disclaimer

The views expressed in this chapter are those of the author and do not necessarily reflect the views or policies of her employer, the U.S. Environmental Protection Agency.

Notes

1 Stratified positionality indicates the differential utility of individuals' and/or groups' wealth, power, and prestige for accessing resources and achieving goals relative to other individuals and groups in specific contexts.
2 Ostrom's fourth level, the meta-constitutional level has similarities with Bhaskar's real level discussed in the section on informal institutions.
3 The term 'social classes' or 'social groups' (Bourdieu, 1990) helps us to avoid the normative understandings of communities as locally, bounded, homogenous social entities constituted of people with shared histories, experiences, and interests (Agrawal and Gibson,1999, Collins, 2010, Leach, Mearns and Scoones, 1997). Grouping people based on common classification schemes for races or ethnicities is also a problem (Brubaker, 2002). Groups identified by sex, race, ethnicity, or any other *single* measure of diversity are frequently extremely diverse. Comparatives should be based on similarities in resources (social, cultural, economic capital) and life chances.
4 South Dakota is located on the ancestral, traditional, and contemporary lands of the Oceti Sakowin – Seven Council Fires – which includes the Santee-Dakota, Yankton-Nakota, and Teton-Lakota. The South Dakota PPR, which constitutes the entire eastern region of South Dakota, is home to 21% of the Native population and four of nine federally recognized Native Nations.
5 Hutterites are an ethnoreligious communal group of Anabaptists who, like the Amish and Mennonites, trace their roots to the Radical Reformation of the early 16th century. On average, each colony consists of 80 people in 14 families who hold all material goods in common and rely on farming and light manufacturing. South Dakota is home to 46% of the United States Hutterite population (May, 2021a).
6 Until 2013, blue crab regulations focused on size limits, locations, and harvest methods and equipment specifications.
7 Quantity of fish harvested, as well as size and sex, as reported by fish houses where fishers sell their catch is a primary means of assessing fish population status.

References

Abidi-Habib, M. and Lawrence, A. (2007) Revolt and remember: How the Shimshal Nature Trust develops and sustains social-ecological resilience in northern Pakistan, *Ecology and Society*, 12(2): 35. https://www.ecologyandsociety.org/vol12/iss2/art35/

Agrawal, A. and Gibson, C.C. (1999) Enchantment and disenchantment: The role of community in natural resource conservation, *World Development*, 27(4): 629–649. https://doi.org/10.1016/S0305-750X(98)00161-2

Archer, M.S. (1995) *Realist Social Theory*, Cambridge: Cambridge University Press.

Bhaskar, R. (2008) *A Realist Theory of Science*, London and New York: Routledge.

Bourdieu, P. (1990) *The Logic of Practice*, translated by R. Nice, Stanford, CA: Stanford University Press.

Brubaker, R. (2002) Ethnicity without groups, *European Journal of Sociology*, 43(2): 163–189. https://doi.org/10.1017/S0003975602001066

Calderon-Contreras, R. and White, C.S. (2020) Access as the means for understanding social-ecological resilience: Bridging analytical frameworks, *Society & Natural Resources*, 33(2):205–223. https://doi.org/10.1080/08941920.2019.1597233

Collins, P.H. (2010) The new politics of community, *American Sociological Review*, 75(1): 7–30. https://doi.org/10.1177/0003122410363293

Crawford, S.E.S. and Ostrom, E. (1995) A grammar of institutions, *American Political Science Review*, 89(3): 582–600. https://doi.org/10.2307/2082975

de Kraker, J. (2017) Social learning for resilience in social–Ecological systems, *Current Opinion in Environmental Sustainability*, 28: 100–107. https://doi.org/10.1016/j.cosust.2017.09.002

Fernández-Giménez, M.E., Augustine, D.J., Porensky, L.M., Wilmer, H., Derner, J.D., Briske, D.D. and Olsgard Stewart, M. (2019) Complexity fosters learning in collaborative adaptive management, *Ecology and Society*, 24(2): 29. https://doi.org/10.5751/ES-10963-240229

Flatschart, E. (2017) Feminist standpoints and critical realism. The contested materiality of difference in intersectionality and new materialism, *Journal of Critical Realism*, 16(3): 284–302. https://doi.org/10.1080/14767430.2017.1313650

Giddens, A. (1984) *The Constitution of Society: Outline of the Theory of Structuration*, Berkeley, CA: University of California Press.

Giddens, A. (1990) *The Consequences of Modernity*, Stanford: Stanford University Press.

Gunderson, L. H. and Holling, C. S. (eds.) (2002) *Panarchy: Understanding Transformations in Human and Natural Systems*, Washington, DC: Island Press.

Leach, M., Mearns, R. and Scoones, I. (1997) Challenges to community based sustainable development – dynamics, entitlements, institutions, *IDS Bulletin-Institute of Development Studies*, 28(4): 4–14. https://doi.org/10.1111/j.1759-5436.1997.mp28004002.x

Martinez, A., Martin, L. and Marlow, S. (2014) Developing a critical realist positional approach to intersectionality, *Journal of Critical Realism*, 13(5): 447–466. https://doi.org/10.1179/1476743014Z.00000000043

May, C.K. (2011) *Visibility, Legitimacy, and Power: A North Carolina Fishing Community and Governance of the Commons*, Doctor of Philosophy (Ph.D.), Fort Collins: Department of Sociology, Colorado State University.

May, C.K. (2012) Active non-participation among local natural resource-dependent communities, *Journal of Environmental Management*, 113: 407–416. https://doi.org/10.1016/j.jenvman.2012.09.022

May, C.K. (2016) Visibility and invisibility: Structural, differential, and embedded power in collaborative governance of fisheries, *Society & Natural Resources*, 29(7): 759–774. https://doi.org/10.1080/08941920.2015.1072257

May, C.K. (2021a) Institutional panarchy: Adaptations in socio-hydrological governance of the South Dakota Prairie Pothole Region, USA, *Journal of Environmental Management*, 293: 112851. https://doi.org/10.1016/j.jenvman.2021.112851

May, C.K. (2021b) Coastal community resilience and power: A comparative analysis of adaptability in North Carolina and Louisiana, *Environmental Management*, 68(1): 100–116. https://doi.org/10.1007/s00267-021-01482-x

May, C.K. (2022) Complex adaptive governance systems: A framework to understand institutions, organizations, and people in socio-ecological systems, *Socio-Ecological Practice Research*, 4: 39–54. https://doi.org/10.1007/s42532-021-00101-7

NCDMF. (2020) *North Carolina Blue Crab (Callinectes Sapidus) Fishery Management Plan Amendment 3*, Morehead City, NC: North Carolina Division of Marine Fisheries (NCDMF).

North, D.C. (1990) *Institutions, Institutional Change and Economic Performance*, New York: Cambridge University Press.

Ostrom, E. (2005) *Understanding Institutional Diversity*, Princeton: Princeton University Press.

Perrow, C. (1961) The analysis of goals in complex organizations, *American Sociological Review*, 26(6): 854–866.

Perrow, C. (1986) *Complex Organizations: A Critical Essay*, New York: Random House.

Ray, V. (2019) A theory of racialized organizations, *American Sociological Review*, 84(1): 26–53. https://doi.org/10.1177/0003122418822335

Rothstein, R. (2017) *The Color of Law: A Forgotten History of How Our Government Segregated America*, New York: Liveright Publishing.

Schlager, E. and Ostrom, E. (1992) Property-rights regimes and natural-resources – A conceptual analysis, *Land Economics*, 68(3): 249–262. https://doi.org/10.2307/3146375

Selznick, P. (1949) *Tva and the Grass Roots: A Study in the Sociology of Formal Organization*, Berkeley and Los Angeles: University of California Press.

Taylor, D.E. (2015) Gender and racial diversity in environmental organizations: Uneven accomplishments and cause for concern, *Environmental Justice*, 8(5): 165–180. https://doi.org/10.1089/env.2015.0018

Tschakert, P., Dietrich, K., Tamminga, K., Prins, E., Shaffer, J., Liwenga, E. and Asiedu, A. (2014) Learning and envisioning under climatic uncertainty: An African experience, *Environment and Planning A*, 46(5): 1049–1068. https://doi.org/10.1068/a46257

Turnhout, E., Metze, T., Wyborn, C., Klenk, N. and Louder, E. (2020) The politics of co-production: Participation, power, and transformation, *Current Opinion in Environmental Sustainability*, 42: 15–21. https://doi.org/10.1016/j.cosust.2019.11.009

Winker, G. and Degele, N. (2011) Intersectionality as multi-level analysis: Dealing with social inequality, *European Journal of Women's Studies*, 18(1): 51–66. https://doi.org/10.1177/1350506810386084

Wyborn, C. (2015) Co-productive governance: A relational framework for adaptive governance, *Global Environmental Change*, 30: 56–67. https://doi.org/10.1016/j.gloenvcha.2014.10.009

6 Participatory modeling of institutional dynamics

Navigating epistemological challenges in fuzzy cognitive mapping

Jelto Makris, Michael Kriegl, Ben Nagel and Achim Schlüter

Introduction

Sustainability research widely acknowledges the importance of involving non-scientific stakeholders in research processes. The insights of those directly working with a system on a daily basis are essential for understanding and addressing complex issues in natural resource governance (Bennett, 2016; Berkes et al., 2000). Participatory modeling has emerged as 'a purposeful learning process for action that engages the implicit and explicit knowledge of stakeholders to create formalized and shared representations of reality' (Voinov et al., 2018: 1). Fuzzy cognitive mapping (FCM) is one participatory modeling method that involves stakeholders to create structured causal maps that represent the dynamics of complex systems. Through qualitative network models with quantitatively weighted relationships, stakeholders can share their perception of how institutions influence natural resource governance, and how they enable or constrain behavioral responses and social-ecological outcomes under system change. FCM has the potential to (1) enhance understanding of institutions and institutional change within social-ecological systems with the help of stakeholder knowledge and (2) leverage this system understanding to facilitate more appropriate interventions in natural resource management. By involving local stakeholders in modeling institutional and social-ecological changes, the method can ultimately contribute to transitioning toward more sustainable and effective institutions.

In the realm of institutional analysis, FCM excels in its ability to integrate and represent a broad spectrum of concepts encompassing non-material social factors, environmental variables, formal and particularly also informal institutions. For studying the latter, participatory methods like FCM are suitable as they unveil informal institutions through the lens of those who engage with them. These actors are the experts of the institutions within the particular context. Moreover, FCM can highlight perceptions of cause-and-effect mechanisms underlying institutional success and failure or disentangle heterogeneous

DOI: 10.4324/9781032624679-6

worldviews regarding a particular regulation or policy (Christen et al., 2015). Getting a grasp of actor's mental models (interpretation how the world works) and ideologies (how it should work) is specifically important when one aims to understand institutional change under uncertain conditions (Denzau and North, 1994), which characterize complex social-ecological systems (Epstein et al., 2020). Substantial differences in mental models and ideologies are important factors impeding successful institutional change (North, 2005).

This chapter presents the use of FCM in a case study on the governance of Harmful Algal Blooms (HABs) in coastal Peru, which revealed potential institutional adaptations. It serves as a specific illustration of FCM's application for institutional analysis, discussing the decisions made and the participatory challenges encountered at the intersection of modeling and stakeholder participation. This chapter emphasizes the importance of acknowledging epistemological challenges, and power issues linked to FCM's design choices from the outset of the research process. This not only helps in navigating challenges throughout the research but can also promote more realistic expectations of the tool's participatory potential for informing institutional change. Through this exploration, we contribute to a more nuanced understanding of FCM by recognizing participation as one among conflicting research goals, complimenting previous existing FCM literature that comprehensively covers other methodological and procedural aspects (Jetter and Kok, 2014; Özesmi and Özesmi, 2004).

This chapter first introduces FCM as a participatory modeling method and discusses its potential uses for institutional analysis. It demonstrates FCM's practical application through a case study. It then examines epistemological and power challenges arising during the case study. Finally, it offers tentative recommendations for future research using FCM for institutional analysis.

Introducing fuzzy cognitive mapping

FCM, initially developed for qualitative social research, expanded to mathematics and computer science in the 1980s and has gained popularity in the last two decades, especially in applied sciences such as business planning, medicine, environmental management and sustainability research (Jetter and Kok, 2014). During FCM, participants identify relevant variables in a system and illustrate their interconnections with causal arrows (Jetter and Kok, 2014). They then assign positive and negative degrees of influence using categorical descriptors (e.g., low-medium-high) or semi-quantitative values ranging from 0 to 1 based on fuzzy logic – which means that relations between variables are based on uncertain and hazy degrees of causality which typically inform people's decision-making (Jetter and Kok, 2014; Kosko, 1986). Figure 6.1 provides an example of an FCM from the case-study presented in this chapter.

The maps are first a static snapshot of a system, complemented by participants' storylines. This aids, for instance, in identifying institutions within a specific resource governance context and understanding the perceived causal mechanisms by which these institutions shape the environment and vice-versa. Being semi-quantitative, FCM can also be used to structurally explore the diversity of

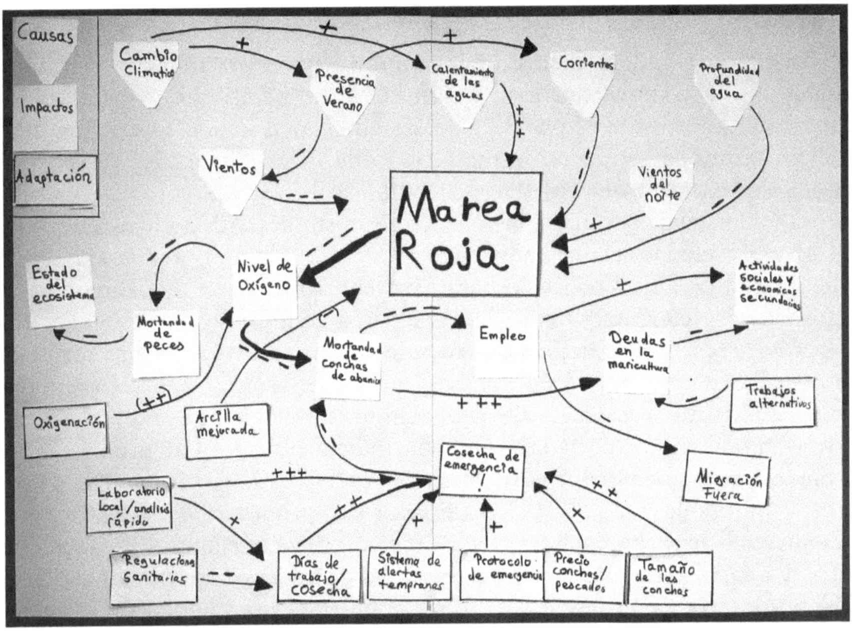

Figure 6.1 Example of a fuzzy cognitive map from the case-study on HABs.

perceptions, ideologies and beliefs related to that system through metrics from network analysis and graph theory (Özesmi and Özesmi, 2004). Furthermore, researchers or stakeholders can investigate the maps using so-called 'what-if' scenarios, increasing or decreasing specific variables to observe feedback dynamics within the system (Özesmi and Özesmi, 2004). These scenarios can help researchers and stakeholders explore and better understand how a given system is predicted to change under varying conditions (Edwards and Kok, 2021). As our mental models influence our decision-making, FCM is also used to help understand the factors guiding the actions and choices of stakeholders.

While many papers emphasize these analytical capabilities of FCM, the maps themselves are the method's key output (Barbrook-Johnson and Penn, 2022). Serving as a form of causal storytelling, they elucidate and visually present to the participants the logical implications of their diverse beliefs about system structure, illustrating which factors are more significantly driven, influenced or caused by external changes or interventions. Moreover, FCM is user-friendly, and intuitive even for researchers and stakeholders without previous modeling experience. It can provide a quick overview of the important factors related to a phenomenon and then facilitate group deliberation and discussion over them (Barbrook-Johnson and Penn, 2022). This makes the tool well suited for explorative research, with limited data and information from prior studies, or when written information related to institutions is not available. Furthermore, the construction of causal maps offers great flexibility in design, allowing for various approaches to be taken (Voinov et al., 2018).

Mapping institutional dynamics with stakeholders

Understanding and modeling of institutional processes influencing system change is a key knowledge gap, which FCM can contribute to by mapping the expert knowledge of people who are embedded in and interacting with these institutional structures in their day-to-day lives. There are two very different approaches to knowledge generation in FCM, which can be used to better understand institutional dynamics. First, studies with positivist roots use FCM to develop objective knowledge about a system and predict real-world phenomena or future policy scenarios by representing and modeling 'expert knowledge' (Jetter and Kok, 2014). Related to this, studies combine the expertise from diverse actors, to leverage the strengths of various forms of knowledge, an idea based on the 'wisdom of the crowd' theory (Aminpour et al., 2020). By mathematically aggregating people's beliefs and viewpoints, 'crowd models' are expected to overcome individual biases and 'produce more complete and potentially more accurate representations of complex problems' (Aminpour et al., 2021: 1). This approach proposes FCM as a time and resource-efficient approach to harness the collective intelligence of people to model a social-ecological system and simulate the impact of potential future institutional interventions in data-scarce contexts (Knox et al., 2023).

Second, and particularly useful for institutional analysis, FCM maps can be explored to compare the diversity in perceptions, mental models, and ideologies of individuals and groups concerning the governance of natural resources (Özesmi and Özesmi, 2004). Within this interpretivist framing, the empirical 'validity' of the participant or crowd's model is less important. Rather, the FCM maps reflect the subjective mental models of people involved in the system which are likely to be more important than reality itself in shaping their actions and decisions, and thereby, institutional change. For example, FCM can help shape more effective institutional arrangements by explicitly modeling what different stakeholders view as the positive and negative impacts of policy, or mapping factors influencing compliance and decision-making across diverse actor groups (Christen et al., 2015; Wade and Biedenweg, 2019). In addition to generating knowledge on existing institutional dynamics, participatory modeling can also inform institutional change by involving stakeholders to co-create models based on broader instrumental, political and normative goals that go beyond the research process itself. For instance, the maps can be used for sharing reflections among researchers and stakeholders to identify areas of conflicting perceptions, while condensing reality in a way that facilitates dialogue and exploration over potential solutions. Here the participatory process is used as a problem-solving tool, for instance to improve management plans and the quality of institutions such as by simulating outcomes of potential policy scenarios (Christen et al., 2015; Wade and Biedenweg, 2019). Another goal of participatory modeling is social learning and the transfer of knowledge as an outcome of stakeholder deliberation or communication (van Vliet, Kok and Veldkamp, 2010). Actors involved and researchers alike might change their understandings and perceptions of the system at stake. FCM research can also aid in fostering shared conceptions of a system by creating and discussing aggregated crowd models, thus reinforcing and highlighting recurring perceptions across the study group (Gray et al., 2019). Mental modeling activities can lead to changes in relationships between stakeholders, and to considerations of longer time horizons when discussing governance decisions (Steger et al., 2021). This process of 'communicative rationality' is crucial for tackling wicked problems in natural resource governance that often have no clear solution and require stakeholders to negotiate conflicting goals to find collective agreement on action (Habermas, 1981/1984; Reed et al., 2009). Nevertheless, stakeholder participation can also have political motivations, arguably promoting trust and dialogue between participants and making the work of research and related policy processes more legitimate (Mielke et al., 2016). Finally, from a normative standpoint, participation is seen as possessing inherent values, fostering democratic practices, accountability, equity, fairness or empowerment (Reed, 2008).

In natural and managerial sciences, where FCM is most frequently applied, these participatory tools are frequently depicted with optimism regarding their goals, especially when seen as having the potential to enhance learning, improve decision-making or empower stakeholders to foster institutional transformations (Tsouvalis and Waterton, 2012; Voinov et al., 2016). Nevertheless, social and political science research has shown that participatory practices are often unable to fulfill their promises due to power issues, limited participation and theoretical and epistemological constraints (Cooke and Kothari, 2001). Moreover, study objectives involve trade-offs, and deeper levels of participation in FCM may not always be desirable given the research goal. For example, when using FCM to understand institutional dynamics, deeper stakeholder participation could limit the analytical depth or deviate from the institutional focus. Conversely, if FCM serves as a tool for reflection and learning to inform locally acceptable institutional change, a deeper and interactive participatory approach becomes essential. Within the same FCM study, goals cannot be equally achieved due to methodological choices and epistemological challenges. Some of these choices and challenges are illustrated through the following case study.

Case study: analysing the social and institutional dimensions of Harmful Algal Blooms in coastal Peru

In this section, we offer a brief overview of the practical steps involved in FCM based on a case study investigating the social and institutional aspects of HABs in coastal Peru. While the main aim of this study was to explore stakeholder perceptions regarding HABs and determine the extent to which adaptive capacities differ between stakeholders, we highlight those choices in research design and challenges in stakeholder participation that are particularly relevant for institutional analysis. For more detailed methodological guidance, readers can refer to other works on FCM (e.g. Gray et al., 2014; Mourhir, 2021; Olazabal et al., 2018; Özesmi and Özesmi, 2004).

HABs, caused by the proliferation of algae and often associated with the appearance of patches of unusual color in the ocean, pose significant risks to both ecosystems and human activities. These algae blooms can trigger the depletion of oxygen in the marine environment as well as the release of toxins, directly impacting marine life as well as human health (Anderson, 2009). HAB-associated closures of coastal areas and fishing grounds significantly affect communities relying on the ocean for fishing, aquaculture or tourism (Ferrol-Schulte et al., 2013). Consequently, HABs have intensified rural poverty, food insecurities and contributed to declines in seafood exports and tourism around the world (Ekstrom et al., 2020; Van Dolah et al., 2016). Despite increasing reports and research on HABs in Peru (Cuellar-Martinez et al., 2023; Cueto-Vega et al., 2022; Orozco et al., 2017), their socio-economic impacts along with governance frameworks and adaptation strategies remain understudied (see Badjeck et al., 2009; Kluger et al., 2022).

To address this lack of understanding in Peru as well as Latin America in general (López-Cortés et al., 2019), we used FCM to investigate stakeholder perceptions on HAB drivers and socio-economic impacts, identifying institutional gaps and highlighting potential adaptation pathways. This knowledge is particularly salient for developing legitimate disaster response strategies (Willis et al., 2018) as well as more adequate institutions to be able to more effectively cope with multi-faceted social-ecological crises (Armstrong et al., 2022; Kriegl et al., 2021; Schwermer et al., 2021).

Study context: the bays of Sechura and Paracas

Sechura Bay in the North and Paracas Bay in the South of Peru are two bay systems influenced by the highly productive Humboldt Current Upwelling System (Garteizgogeascoa et al., 2020). Small-scale fisheries, mariculture as well as tourism are important activities in these coastal areas, which are particularly vulnerable to HABs (Brown et al., 2020; Díaz et al., 2019). Sechura Bay is a significant aquaculture site both within Peru as well as within Latin America, with an important scallop production sector that employs tens of thousands of workers (Kluger et al., 2019). The scallop sector in this bay is characterized by a private property regime with most scallop concessions controlled by private investors (Schlüter et al., 2021). Paracas Bay, with its historical importance for scallop production (Kluger et al., 2019), nowadays hosts a spectrum of activities including a small mariculture sector as well as small-scale and industrial fishing associated with large fish processing plants, alongside a booming tourism industry (Garteizgogeascoa et al., 2020). HABs, marine heatwaves and the effects of El Niño events greatly affect scallop cultivation (Cuellar-Martinez et al., 2023; Cueto-Vega et al., 2022; Kluger et al., 2022; Kriegl et al., 2021, 2022; Sánchez et al., 2022).

Methods: conducting FCM in practice

In late 2022, we used FCM to collect mental models around the concept of HABs during 18 in-person focus groups. Relevant stakeholder groups included small-scale fishers, aquaculture operators (both small- and large-scale), scientists as well as representatives of local authorities. These groups were identified based on literature and existing knowledge of the system (Badjeck et al., 2009; Kluger et al., 2022; Kriegl et al., 2022; Orozco et al., 2017). Study participants were selected using snowball sampling with various entry points into the system (Jetter and Kok, 2014).

In small focus groups (<6 people from within the same stakeholder group; Nyumba et al., 2018), participants were invited to brainstorm relevant concepts based on guiding questions related to (a) drivers, (b) impacts and (c) adaptation strategies in the context of HABs as well as to indicate relevant connections (influence arrows) between them. In a second step, participants then assigned the strength for each identified causal connection (+++, ++, +, -, --, ---). The maps were created manually using portable whiteboards, markers and post-its (facilitating discussions in various interview settings and places but making manual post-fieldwork digitization necessary) and sessions were recorded with participant consent to better capture the participants' explanation of the system.

In FCM studies, methodological choices fundamentally influence the whole research process. Key questions relevant for this study are thus outlined in Table 6.1, drawing on guidance from experts such as Özesmi and Özesmi (2004), Jetter and Kok (2014), and Gray et al. (2014). To address concerns about transparency, a reporting procedure by Olazabal et al. (2018) was adopted.

Data analysis

After completing all focus groups, the analysis comprised several steps. Initially, qualitative data from focus groups was analyzed to gain a detailed understanding of stakeholders' perspectives on system components and the social and institutional context of HABs in coastal Peru. Participant definitions and explanations of mentioned concepts were systematically organized into a codebook. Next, 'raw' maps were pre-processed to establish a common terminology and enhance map coherence (Olazabal et al., 2018). This stage involved standardizing concept names, correcting errors and merging closely related variables (Jetter and Kok, 2014). It is usually the stage where the researcher influences the representation of the stakeholder's knowledge most. To minimize researcher influence and preserve participants' representations of the system, the maps were processed as little as possible. The processed maps were then digitized and integrated into the mental modeler software (www.mentalmodeler.com; for an overview of different software options see Felix et al., 2019).

Table 6.1 Procedural steps to be followed for a reflected use of FCM

Stage 1. Scope & objectives
What is the research goal?
What are the relevant system boundaries?
What is the final model output? Focus on heterogeneity of stakeholder perception (individual maps or aggregated per stakeholder group) or 'crowd wisdom' approach (community map) or a combination of both?
Stage 2. Data collection
How to choose participants? How to define stakeholder groups?
Map setting: individual interviews, focus groups or both?
What meaning for the map connections/arrows: dynamic or causal?
What weights to assign to the causal links?
Composition of participants during focus groups – mixed groups or not?
What place? What length for the focus group discussion?
Workshop approach: pre-defined concepts, free drawing, or guiding questions?
What role for the researcher during discussion?
How many maps are needed?
Stage 3. Data processing
How to code, categorize the qualitative data/concept meanings?
How much qualitative pre-processing of the maps?
What method for aggregation?
What software to use for the processing and analysis?
Stage 4. Data analysis
Which variables to focus on during the scenario analysis?
What network metrics to choose?
What activation function for the scenario analysis?
How strongly should the variables be activated?
What variables to focus on during the analysis?

Source: Authors' own elaboration drawing on Özesmi and Özesmi (2004), Jetter and Kok (2014), and Gray et al. (2014).

Subsequently, the maps were aggregated based on pre-defined stakeholder groups to create different community models and analyzed to explore perceptions of HABs' impacts and drivers. For this, the individual maps (i.e., the underlying adjacency matrices; Özesmi and Özesmi, 2004) were overlaid using a software routine (Aminpour, 2019), highlighting shared beliefs across individual maps. The community models for the five stakeholder groups were compared, and a final 'crowd' model was developed by aggregating all stakeholder maps. Different network metrics were calculated to describe the aggregated maps in terms of network structure (Singh and Chudasama, 2017).

Finally, the scenario analysis tool of the mental modeler software ('What if …?') was used on the aggregated community map to assess the system's response to simulated increases in HABs and evaluate the effectiveness of adaptation strategies and institutional responses for Peru's small-scale and large-scale scallop mariculture sector. This involved 'activating' variables in the system and observing their influence on related concepts. It is essential to note that this scenario analysis represents participant perceptions, not statistical predictions (Singh and Chudasama, 2017). A baseline scenario ('no adaptation') and different adaptation scenarios were simulated, focusing on the mitigation of HAB impacts on key variables such as scallop mortality, debts, employment and income.

Case study results

The analysis revealed areas of consensus and disagreement on HAB impacts, drivers and adaptation strategies. Firstly, stakeholders agreed on significant ecological, social and economic impacts. Secondly, drivers of HABs were perceived as being mostly out of control of local actors. Mentioned responses to HABs were thus primarily based on adaptation rather than prevention or mitigation. The most significant finding concerning institutional aspects was that the combination of diverse stakeholder knowledge helped to identify 38 adaptation strategies – several of institutional nature. Among these, the implementation of an early warning system and the extension of current monitoring schemes were deemed the most effective. Moreover, while the impacts of HABs on the scallop sector were viewed as substantial throughout, small-scale aquaculture operators were perceived as more susceptible compared to large-scale producers, revealing existing inequalities. In this context, not all identified adaptation strategies were equally accessible to large-scale and small-scale scallop producers alike, reflecting institutional barriers. Factors such as information asymmetry (with large companies having better access to information and technological solutions) as well as the distribution of access rights in the bay and the processing sector influenced adaptive capacity.

Using FCM in this context represents a tentative step toward identifying solutions for HABs and addressing inequalities in adaptive capacities among different stakeholders. This underscores FCM's value in understanding institutions related to natural resource governance, for highlighting institutional gaps and barriers as well as for proposing improvements based on stakeholder priorities. Nonetheless, challenges related to the participatory process of FCM emerged, including stakeholder participation depth, power dynamics and epistemological considerations, which will be discussed in the next section.

Discussion

Level of participation in fuzzy cognitive mapping for institutional analysis.

There are many different levels of stakeholder participation possible when conducting an FCM study. Voinov and Bousquet (2010) outline stages in which stakeholders can be involved in participatory modeling including goal setting, stakeholder analysis, design of modeling, building the models, processing them, defining scenarios and analyzing and evaluating them. If the aim is to inform governance interventions, stakeholders should ideally be engaged interactively and iteratively, facilitating a two-way flow of information between stakeholders and researchers from the very beginning to the end (Voinov and Bousquet, 2010). As discussed, deeper participation could enhance local context suitability, potentially increasing trust, legitimacy, acceptance and uptake. However, when the goal is to improve the understanding of why and how institutions change, deeper stakeholder participation in FCM – for instance, during problem definition – may not lead to a focus on institutional or governance aspects.

In this case study, the topic of HABs was chosen by the researchers, as it emerged as a topic of importance in previous research. Nevertheless, to increase the influence of local experts, the FCM focus groups were designed with minimal researcher intervention, allowing participants to freely choose concepts. Consequently, adaptation strategies were entirely determined by participants. With researchers' intention not to influence the topics addressed, around 18 of 38 suggested adaptation measures were of an institutional nature, while significant discussion time was also dedicated to other non-institutional issues. An open mapping process is a good approach, for instance to find out if institutional issues are perceived as a significant problem for governance in the first place. For a detailed understanding of institutional dynamics, however, the initial discussion question could have been more focused on institutional aspects (i.e., narrow system boundaries). As an additional step, pre-defined concepts directly relating to institutional analysis could have been provided, for example, by drawing on theoretical frameworks such as the social-ecological system framework. Pre-defining questions and concepts, either developed by researchers or through preliminary surveys or workshops, can ease the brainstorming process and reduce recall error for respondents, but it comes with the trade-off of heightened researcher influence and reducing participants' freedom in concept selection. Pre-defining concepts is often also done in FCM studies for making the building of aggregated models easier, as the process of FCM homogenization is time-consuming and typically based on the subjective judgment of the researcher. The lack of participation at this processing stage might lead to the perception of 'black-boxing' the model construction.

Finally, the models created allow for participatory exploration and discussion of various jointly identified scenarios. However, such participatory process requires extensive time and funding. The current constraints of the scientific process, such as short project cycles, limited fieldwork funding, the structure of academic knowledge products (articles, thesis) and the still valid traditional institutional incentives within the science system, focusing very much on highly ranked academic journal articles often lead to a reduced use of those participatory discussions, as has been the case in this study.

Power and representation

Power dynamics present another significant challenge, particularly when the goal is to facilitate more appropriate intervention in natural resource governance. If certain groups are omitted from the research process, the institutional hurdles they face may go unnoticed. In the pursuit of building better institutions, neglecting crucial interest groups could yield unintended consequences. There is a risk that the results reinforce the interests of privileged groups, marginalizing others and seemingly validating their claims under participatory rhetoric (Cornwall, 2008; Stringer et al., 2006). In FCM, where the final products typically represent 'group' or 'community' models, careful consideration of power relations within and between these groups is essential. This is also to avoid framing the stakeholders as a homogenous social structure, failing to consider internal differences and political processes (Agrawal and Gibson, 1999).

Particularly if FCMs are built in groups, as was the case in this study, factors such as gender dynamics, education, argumentative skills, language barriers, socio-economic and formal status and other social factors can lead to research processes being dominated by some (Lang et al., 2012; Stringer et al., 2006). This phenomenon is important to consider for any empirical study, but it is particularly crucial for FCM, which rely on and require a specific logic that may privilege one actor over another. In the study, for instance, participants who quickly understood the map structure and causal logic contributed more. To address this, variables and connections were drawn based on explanations from participants who were less inclined to edit the map directly. Nevertheless, this heightened the researchers' influence on the model, creating a trade-off between inclusivity and researcher bias.

Finally, the issue of whose perception is represented extends to the analysis stage, given that choices need to be made on how maps are processed and aggregated. For instance, opting for a map aggregation method favoring shared knowledge can lead to final models emphasizing consensus while downplaying minority views and divergence in perception. Together, these challenges underscore that group models do not neutrally represent diverse stakeholder views, as 'ideal speech situations' (Habermas, 1981/1984) during group discussions are rare, and aggregation methods prioritize certain aspects over others. The omission of particular views may not pose a significant issue, when the study aims to analyze and understand an institutional system. However, if the aim is to explore new institutional solutions, marginalizing certain perspectives can be detrimental.

Epistemological issues

FCM translates qualitative storylines into semi-quantitative models and then models back into stories (Beck, 2018). As such, FCM comes with underlying values and epistemological assumptions that shape the type of information that participants express regarding the role of institutions in natural resource governance. Scott (1998) criticizes the oversimplification of local practical knowledge through abstractions such as maps and models, arguing that they strip away detailed, time- and context-specific characteristics, subtle nuances and practical applications that make such knowledge valuable. Scott (1998) highlights that these abstractions have the power and intention to reshape the reality they represent. However, by overlooking facets of the complex existing social order, they potentially lead to adverse outcomes when solutions are derived from them (Agrawal, 2002; Scott, 1998). This valid critique is crucial to consider, particularly when FCMs are used for institution building. It is important to reflect which type of knowledge regarding institutions is emphasized by these tools and how to mitigate the potential loss of the richness of information provided by those directly working in the system.

In the case study, for instance, the pre-determined causal and semi-quantitative frame of FCM caused unease among some participants. Several participants hesitated to assign weights. This led to a proactive role of the researcher in adding causal relations, as participants primarily confirmed or negated questions asked on the strength of the relationship. Participants also struggled to include concepts in the map that they considered essential but lacked a straightforward quantitative relationship with other variables in the system, often resulting in their exclusion from the analysis due to disagreements among participants. The latter issue is of particular concern, if the FCM is done in groups and not individually. Moreover, the logic of mathematical aggregation is limiting the expression of perception differences or disagreement. Conflicting views, such as one participant perceiving a positive connection between two variables while another sees it negatively, may either cancel each other out during map aggregation or the dominant viewpoint – somewhat weakened – enters the final model (Jetter and Kok, 2014). While in theory any factor can be added to the map, these practical issues created a bias towards information that was easily quantifiable and straightforward to define.

This has significant implications for the analysis of institutional aspects, which often do not follow the linear logic inherent to FCM. Instead, non-linear relationships, such as necessary conditions or critical thresholds, are common. In principle, mental models can and do entail non-linear relationships. Nevertheless, current FCM tools are limited to simplified relationships. Another related issue concerns how well FCM design facilitates the analysis of formal and informal institutions. As FCM relies entirely on the knowledge of involved actors, it is particularly suitable for analyzing informal institutions. However, sensitive informal institutions, like bribery, or corruption might be better explored with individual mapping.

To mitigate some of the challenges, discussions could begin as brainstorming sessions focusing on conceptual explanations of relationships, delaying quantification until later in the process. Additionally, the loss in detail and the limitations to linear and simplistic logics can be partially remedied by later analyzing the entire spoken dialogue in detail. This is a source of 'thick' information by capturing stakeholder reflections that is minimally influenced by the researcher, who only asked clarification questions. Moreover, conducting FCMs with individuals instead of focus groups can also alleviate some of the epistemological and power issues, and potentially lead to more detailed and diverse mental models of the institutional structure. However, group modeling may better be suited as a deliberative tool to foster consensus over acceptable solutions and institution building and learning (for further discussion on individual vs. group mapping refer to Gray et al., 2014).

FCMs – especially heterogenous individual ones – require thorough processing to facilitate aggregation, metric analysis and scenario assessment. These tools are crucial to point out differences in mental models among stakeholders (or stakeholder groups), which might complicate institutional change processes. This processing and aggregation, however, can inadvertently strip away nuanced, context-specific information on formal and informal institutions provided by stakeholders, potentially amplifying researcher bias and leading to models that mismatch stakeholders' initial perceptions. Refraining from extensive processing can help. In this study, for instance, despite efforts to minimize influence by condensing and renaming variables as little as possible only based on audio transcriptions, the aggregated maps still deviated from the original data. This underscores the previously discussed importance of deeper participatory processes. Validation and evaluation of the model by stakeholders become crucial, along with creating transparency in processing and analyzing the maps. While the study followed guidelines for transparency and reproducibility useful to other researchers (see Olazabal et al., 2018), these may not necessarily ensure transparency or a thorough understanding for the stakeholders involved, presenting another key challenge.

Finally, as any model is a simplification of reality, trade-offs to alternative methods need to be considered. Despite the case study bias toward more easily quantifiable information, in comparison to other scientific modeling approaches, FCM is still much more open regarding the combinations of variables that can be incorporated, such as concepts relating to institutional dynamics. The only condition is that these can be related in 'fuzzy' causal terms. Compared to many modeling methods, FCM is also more time-efficient, for example, an entire system can be modeled, and potential future policy scenarios simulated and discussed within the span of a single workshop.

Conclusion and recommendations

FCM is gaining increasing popularity as a participatory modeling tool in environmental management and social-ecological system research (Jetter and Kok, 2014; Mourhir, 2021; Papageorgiou and Salmeron, 2013). This chapter aims to illustrate the possible value of FCMs for researching institutions and facilitating institutional change while revealing trade-offs and epistemological challenges related to stakeholder participation and modeling that may affect its efficacy for institutional analysis. Drawing from reflections on the case study, some lessons-learned have emerged. These lessons aim to foster the development of best practices for researchers intending to utilize FCM as a participatory tool for studying institutions in natural resource governance.

1 Trade-offs arise when using FCM either as a participatory tool to inform institutional change or as an analytical tool to enhance understanding of institutions and institutional change. Deeper levels of stakeholder participation may be required if FCM is utilized to facilitate more appropriate interventions in natural resource management. However, involving stakeholders extensively at various research stages might constrain the method's potential for in-depth analysis and comprehension of institutional dynamics.
2 Reflect on the underlying epistemological, ontological, and theoretical assumptions of FCM and their implications for institutional analysis.
3 The case study highlights how underlying assumptions shape which types of knowledge are prioritized during participatory processes, impacting institutional analysis. It is crucial to analyze and acknowledge upfront which 'forms of knowledge' might be marginalized and which stakeholders feel comfortable expressing their views within the specific setup of the concrete FCM study. There are many different forms of setup (individual or group exercises, providing concepts or leaving mapping freely to participants, etc.; refer to Table 6.1). Reflecting on epistemological assumptions can help decide on FCM's suitability for a given context and its ability to foster equitable dialogue among diverse forms of knowledge (Santos, 2014). Nonetheless, the exclusive focus on linear relationships within FCM (e.g., not allowing, e.g., for necessary conditions or thresholds) remains a limitation to understanding institutional and broader social-ecological systems. This limitation should be acknowledged when opting for FCM to analyze institutions.
4 Use FCM to highlight the diversity of perceptions and visually represent the social-ecological complexity of a given context. Despite models aiming to simplify complex realities, FCM holds significant potential as a time-efficient tool for visualizing and elucidating the intricate details of a study context and the varied mental models of diverse actors, particularly when complemented by other qualitative research methods. Understanding the diversity and potential contradictions among different mental models is crucial for comprehending the process of institutional change. This understanding can then inform governance changes if this is the research objective.
5 FCM as a method relying on oral information is particularly suitable for understanding institutions in data-scarce contexts. In contexts where abundant written information exists, analyzing documents (e.g., newspaper articles or Twitter chats) can aid in understanding institutional change. However, where written information is lacking, FCMs offer a quick means of capturing diverse actor perspectives, as well as on informal institutions which might be less documented.
6 Acknowledge power dynamics that determine who participates and whose voices dominate the final models. Power dynamics during FCM shape whose perspectives dominate and which types of knowledge are prioritized, impacting the interventions that potentially emerge from the research process. Therefore, research that uses FCM as a participatory tool for

institution-building needs to critically reflect on tacit power issues that influence the process (Fritz and Binder, 2020).

7 FCMs can aid in understanding institutional choice but may not necessarily pinpoint sustainable institutional solutions. Analyzing the different mental models of various actor groups and assessing their power resources during the institutional bargaining process can help understand which institutional solutions are most likely to be implemented (Knight and North, 2004; Mahoney and Thelen, 2009). This was observed in previous processes of institutional change in the case study (Schlüter et al., 2023). However, these processes also revealed that the changes made did not always result in effective or long-lasting sustainable institutional solutions, as they often overlooked the natural or socio-economic materialities of the system (Kluger et al., 2022).

References

Agrawal, A. (2002) Indigenous knowledge and the politics of classification, *International Social Science Journal*, 54(173): 287–297. https://doi.org/10.1111/1468-2451.00382

Agrawal, A. and Gibson, C.C. (1999) Enchantment and disenchantment: The role of community in natural resource conservation, *World Development*, 27(4): 629–649. https://doi.org/10.1016/S0305-750X(98)00161-2

Aminpour, P. (2019) *PyFCM (Python for Fuzzy Cognitive Mapping)*. Available at: https://github.com/payamaminpour/PyFCM/wiki. Accessed on March 15, 2023.

Aminpour, P., Gray, S.A., Jetter, A.J., Introne, J.E., Singer, A. and Arlinghaus, R. (2020) Wisdom of stakeholder crowds in complex social-ecological systems, *Nature Sustainability*, 3(3): 191–199. https://doi.org/10.1038/s41893-019-0467-z

Aminpour, P., Gray, S.A., Singer, A., Scyphers, S.B., Jetter, A.J., Jordan, R., Murphy, R. and Grabowski, J.H. (2021) The diversity bonus in pooling local knowledge about complex problems. *Proceedings of the National Academy of Sciences*, 118(5): e2016887118. https://doi.org/10.1073/pnas.2016887118

Anderson, D.M. (2009) Approaches to monitoring, control and management of harmful algal blooms (HABs). *Ocean & Coastal Management*, 52(7): 342–347. https://doi.org/10.1016/j.ocecoaman.2009.04.006

Armstrong, A., Stedman, R.C., Sweet, S. and Hairston, N. (2022) What causes harmful algal blooms? A case study of causal attributions and conflict in a Lakeshore community, *Environmental Management*, 69(3): 588–599. https://doi.org/10.1007/s00267-021-01581-9

Badjeck, M.-C., Mendo, J., Wolff, M. and Lange, H. (2009) Climate variability and the Peruvian scallop fishery: The role of formal institutions in resilience building, *Climatic Change*, 94(1–2): 211–232. https://doi.org/10.1007/s10584-009-9545-y

Barbrook-Johnson, P. and Penn, A.S. (2022) Fuzzy cognitive mapping, in Barbrook-Johnson, P. and Penn, A. S. (eds.) *Systems Mapping: How to Build and Use Causal Models of Systems*, Cham: Springer International Publishing. https://doi.org/10.1007/978-3-031-01919-7_6

Beck, M. (2018) Telling stories with models and making policy with stories: An exploration, *Climate Policy*, 18(7), 928–941. https://doi.org/10.1080/14693062.2017.1404439

Bennett, N.J. (2016) Using perceptions as evidence to improve conservation and environmental management, *Conservation Biology*, 30(3): 582–592. https://doi.org/10.1111/cobi.12681

Berkes, F., Colding, J. and Folke, C. (2000) Rediscovery of traditional ecological knowledge as adaptive management. *Ecological Applications*, 10(5): 1251–1262. https://doi.org/10.2307/2641280

Brown, A.R., Lilley, M., Shutler, J., Lowe, C., Artioli, Y., Torres, R., Berdalet, E. and Tyler, C.R. (2020) Assessing risks and mitigating impacts of harmful algal blooms on mariculture and marine fisheries, *Reviews in Aquaculture*, 12(3): 1663–1688. https://doi.org/10.1111/raq.12403

Christen, B., Kjeldsen, C., Dalgaard, T. and Martin-Ortega, J. (2015). Can fuzzy cognitive mapping help in agricultural policy design and communication? *Land Use Policy*, 45: 64–75. https://doi.org/10.1016/j.landusepol.2015.01.001

Cooke, B. and Kothari, U. (2001) *Participation: The New Tyranny?* London/New York: Zed Books.

Cornwall, A. (2008) Unpacking 'participation': Models, meanings and practices, *Community Development Journal*, 43(3): 269–283. https://doi.org/10.1093/cdj/bsn010

Cuellar-Martinez, T., Huanca Ochoa, A. del R., Sánchez, S., Aguirre-Velarde, A., Martínez Ocas, E.O., Rodriguez Velasquez, A.M., Saavedra Querevalú, R.I., Colas, F., Tam, J. and Gutiérrez, D. (2023) Abundance and distribution of potentially toxic phytoplankton in aquaculture sites along the Peruvian coast, *Journal of Marine Systems*, 240: 103865. https://doi.org/10.1016/j.jmarsys.2023.103865

Cueto-Vega, R., Flye-Sainte-Marie, J., Aguirre-Velarde, A., Jean, F., Gil-Kodaka, P. and Thouzeau, G. (2022) Size-based survival of cultured Argopecten purpuratus (L, 1819) under severe hypoxia, *Journal of the World Aquaculture Society*, 53(1): 151–173. https://doi.org/10.1111/jwas.12777

Denzau, A.T. and North, D.C. (1994) Shared mental models: Ideologies and institutions, *Kyklos*, 47(1): 1–31.

Díaz, P., Alvarez, G., Varela, D., Santos, I.E., Diaz, M., Molinet, C., Seguel, M., Aguilera-Belmonte, A., Guzmán, L., Uribe, E., Rengel, J., Hernández, C., Segura, C. and Figueroa, R. (2019) Impacts of harmful algal blooms on the aquaculture industry: Chile as a case study, *Perspectives in Phycology*, 6(1–2): 39–50. https://doi.org/10.1127/pip/2019/0081

Edwards, G.I. and Kok, K. (2021) Building a Fuzzy Cognitive Map from stakeholder knowledge: An Episodic, asynchronous approach, *Current Research in Environmental Sustainability*, 3: 100053. https://doi.org/10.1016/j.crsust.2021.100053

Ekstrom, J.A., Moore, S.K. and Klinger, T. (2020) Examining harmful algal blooms through a disaster risk management lens: A case study of the 2015 U.S. West Coast domoic acid event, *Harmful Algae*, 94: 101740. https://doi.org/10.1016/j.hal.2020.101740

Epstein, G., Morrison, T.H., Lien, A., Gurney, G.G., Cole, D.H., Delaroche, M., Villamayor Tomas, S., Ban, N. and Cox, M. (2020) Advances in understanding the evolution of institutions in complex social-ecological systems, *Current Opinion in Environmental Sustainability*, 44: 58–66. https://doi.org/10.1016/j.cosust.2020.06.002

Felix, G., Nápoles, G., Falcon, R., Froelich, W., Vanhoof, K. and Bello, R. (2019) A review on methods and software for fuzzy cognitive maps, *Artificial Intelligence Review*, 52(3): 1707–1737. https://doi.org/10.1007/s10462-017-9575-1

Ferrol-Schulte, D., Wolff, M., Ferse, S. and Glaser, M. (2013) Sustainable Livelihoods Approach in tropical coastal and marine social–ecological systems: A review, *Marine Policy*, 42: 253–258. https://doi.org/10.1016/j.marpol.2013.03.007

Fritz, L. and Binder, C.R. (2020) Whose knowledge, whose values? An empirical analysis of power in transdisciplinary sustainability research, *European Journal of Futures Research*, 8(1): 3. https://doi.org/10.1186/s40309-020-0161-4

Garteizgogeascoa, M., Kluger, L.C., Gonzales, I.E., Damonte, G. and Flitner, M. (2020) Contextualizing Scenarios to Explore Social-Ecological Futures: A Three Step Participatory Case Study for the Humboldt Current Upwelling System, *Frontiers in Marine Science*, 7: 557181. https://www.frontiersin.org/articles/10.3389/fmars.2020.557181

Gray, S.R.J., O'Mahony, C., O'Dwyer, B., Gray, S.A. and Gault, J. (2019) Caught by the fuzz: Using FCM to prevent coastal adaptation stakeholders from fleeing the scene, *Marine Policy*, 109: 103688. https://doi.org/10.1016/j.marpol.2019.103688

Gray, S.A., Zanre, E. and Gray, S.R.J. (2014) Fuzzy cognitive maps as representations of mental models and group beliefs, in Papageorgiou, E. I. (ed.) *Fuzzy Cognitive Maps for Applied Sciences and Engineering*, Berlin/Heidelberg: Springer. pp. 29–48.

Habermas, J. (1984) *The Theory of Communicative Action. Vol. I: Reason and the Rationalization of Society* (T. McCarthy, Trans.; Vol. 1). Boston, MA: Beacon Press. (Original work published 1981)

Jetter, A.J. and Kok, K. (2014) Fuzzy Cognitive Maps for futures studies – A methodological assessment of concepts and methods, *Futures*, 61: 45–57. https://doi.org/10.1016/j.futures.2014.05.002

Kluger, L.C., Schlüter, A., Garteizgogeascoa, M. and Damonte, G. (2022) Materialities, discourses and governance: Scallop culture in Sechura, Peru, *Journal of Environmental Policy & Planning*, 24(3): 309–324. https://doi.org/10.1080/1523908X.2022.2047620

Kluger, L.C., Taylor, M.H., Wolff, M., Stotz, W. and Mendo, J. (2019) From an open-access fishery to a regulated aquaculture business: The case of the most important Latin American bay scallop (*Argopecten purpuratus*), *Reviews in Aquaculture*, 11(1): 187–203. https://doi.org/10.1111/raq.12234

Knight, J. and North, D. (2004) Explaining Economic Change: The Interplay between Cognition and Institutions, in Menard, C. (ed.) *Controversies and Challenges in the New Institutional Economics* (Vol. 7), Cheltenham: Edward Elgar. pp. 471–486. https://www.redi-bw.de/db/ebsco.php/search.ebscohost.com/login.aspx?direct=true&db=ecn&AN=0827168&site=ehost-live

Knox, C., Gray, S., Zareei, M., Wentworth, C., Aminpour, P., Wallace, R.V., Hodbod, J. and Brugnone, N. (2023) Modeling complex problems by harnessing the collective intelligence of local experts: New approaches in fuzzy cognitive mapping, *Collective Intelligence*, 2(4): 26339137231203582. https://doi.org/10.1177/26339137231203582

Kosko, B. (1986) Fuzzy cognitive maps. International Journal of Man-Machine Studies, 24(1): 65–75. https://doi.org/10.1016/S0020-7373(86)80040-2

Kriegl, M., Kluger, L.C., Gorris, P. and Kochalski, S. (2022) Coastal livelihood resilience to abrupt environmental change: The role of social capital in a Peruvian bay, *Regional Environmental Change*, 22(3): 103. https://doi.org/10.1007/s10113-022-01959-3

Kriegl, M., Kluger, L.C., Holzkämper, E., Nagel, B., Kochalski, S. and Gorris, P. (2021) How Important are Social Networks in Times of Environmental Crises? *Easy Social Sciences*, 66: 11–22. https://doi.org/10.15464/easy.2021.002

Lang, D.J., Wiek, A., Bergmann, M., Stauffacher, M., Martens, P., Moll, P., Swilling, M. and Thomas, C.J. (2012) Transdisciplinary research in sustainability science: Practice, principles, and challenges, *Sustainability Science*, 7(1): 25–43. https://doi.org/10.1007/s11625-011-0149-x

López-Cortés, D.J., Núñez Vázquez, E.J., Dorantes-Aranda, J.J., Band-Schmidt, C.J., Hernández-Sandoval, F.E., Bustillos-Guzmán, J.J., Leyva-Valencia, I. and Fernández-Herrera, L.J. (2019) The state of knowledge of harmful algal blooms of Margalefidinium polykrikoides (aka Cochlodinium polykrikoides) in Latin America, *Frontiers in Marine Science*, 6: 463. doi: 10.3389/fmars.2019.00463

Mahoney, J. and Thelen, K. (2009). *Explaining Institutional Change: Ambiguity, Agency, and Power*, Cambridge: Cambridge University Press.

Mielke, J., Vermaßen, H., Ellenbeck, S., Fernandez Milan, B. and Jaeger, C. (2016) Stakeholder involvement in sustainability science – A critical view, *Energy Research & Social Science*, 17: 71–81. https://doi.org/10.1016/j.erss.2016.04.001

Mourhir, A. (2021) Scoping review of the potentials of fuzzy cognitive maps as a modeling approach for integrated environmental assessment and management, *Environmental Modelling & Software*, 135: 104891. https://doi.org/10.1016/j.envsoft.2020.104891

North, D.C. (2005) *Understanding the Process of Economic Change*, Princeton, NJ: Princeton University Press.

Nyumba, T., Wilson, K., Derrick, C.J. and Mukherjee, N. (2018) The use of focus group discussion methodology: Insights from two decades of application in conservation, *Methods in Ecology and Evolution*, 9(1): 20–32. https://doi.org/10.1111/2041-210X.12860

Olazabal, M., Neumann, M.B., Foudi, S. and Chiabai, A. (2018) Transparency and Reproducibility in Participatory Systems Modelling: The Case of Fuzzy Cognitive Mapping, *Systems Research and Behavioral Science*, 35(6): 791–810. https://doi.org/10.1002/sres.2519

Orozco, R., Quispe, Y., Lorenzo, A. and Zamudio, M.L. (2017) Asociación de floraciones de algas nocivas y Vibrio spp. En áreas de pesca y acuicultura de bivalvos de moluscos en las bahías de Sechura y Pisco, Perú, *Revista Peruana de Biología*, 24(1): 111–116.

Özesmi, U. and Özesmi, S.L. (2004) Ecological models based on people's knowledge: A multi-step fuzzy cognitive mapping approach, Ecological Modelling, 176(1-2): 43–64. https://doi.org/10.1016/j.ecolmodel.2003.10.027

Papageorgiou, E.I., and Salmeron, J.L. (2013) A review of fuzzy cognitive maps research during the last decade, *IEEE Transactions on Fuzzy Systems*, 21(1): 66–79. https://doi.org/10.1109/TFUZZ.2012.2201727

Reed, M.S. (2008) Stakeholder participation for environmental management: A literature review. *Biological Conservation*, 141(10): 2417–2431. https://doi.org/10.1016/j.biocon.2008.07.014

Reed, M.S., Graves, A., Dandy, N., Posthumus, H., Hubacek, K., Morris, J., Prell, C., Quinn, C.H. and Stringer, L.C. (2009) Who's in and why? A typology of stakeholder analysis methods for natural resource management, *Journal of Environmental Management*, 90(5): 1933–1949. https://doi.org/10.1016/j.jenvman.2009.01.001

Sánchez, S., Delgado, E., Bernales, A., Jacobo, N., Franco, A. and Correa, D. (2022) Floraciones algales nocivas en la costa peruana durante El Niño Costero 2017 y su relación con las condiciones ambientales. *Boletin Instituto Del Mar Del Perú*, 36(2): 452–462. https://doi.org/10.53554/boletin.v36i2.347

Santos, B. de S. (2014) *Epistemologies of the South: Justice against Epistemicide* (1st edition), New York: Routledge.

Schlüter, A., Kluger, L.C., Damonte, G. and Garteizgogeascoa, M. (2021) *From securing rights for small-scale fishers to property for "efficient" producers: The privatisation of scallop production in Sechura Bay, Peru,* Social Science Research Network (SSRN). https://ssrn.com/abstract=3994780

Schlüter, A., Kluger, L.C., Garteizgogeascoa, M. and Damonte, G. (2023) Resource grabbing and the blue commons: The evolution of institutions in scallop production in Sechura Bay, Peru, in Neef, A., Ngin, C., Moreda, T. and Mollett, S. (eds.), *Routledge Handbook of Global Land and Resource Grabbing* (1st edition), London: Routledge. pp. 300–317.

Schwermer, H., Aminpour, P., Reza, C., Funk, S., Möllmann, C. and Gray, S.A. (2021) Modeling and understanding social-ecological knowledge diversity, *Conservation Science and Practice, 3*(5): e396. https://doi.org/10.1111/csp2.396

Scott, J.C. (1998) *Seeing Like a State: How Certain Schemes to Improve the Human Condition Have Failed,* New Haven, CT: Yale University Press.

Singh, P.K. and Chudasama, H. (2017) Assessing impacts and community preparedness to cyclones: A fuzzy cognitive mapping approach, *Climatic Change,* 143(3–4): 337–354. https://doi.org/10.1007/s10584-017-2007-z

Steger, C., Gebrehiwot, K., Chengere, S.A., Marinkovich, J., Dullo, B.W., Zewde, S.W. and Klein, J.A. (2021) Mental models of a social-ecological system facilitate social learning among a diverse management team, *Environmental Science & Policy,* 122: 127–138. https://doi.org/10.1016/j.envsci.2021.04.006

Stringer, L.C., Dougill, A.J., Fraser, E., Hubacek, K., Prell, C. and Reed, M.S. (2006) Unpacking "participation" in the adaptive management of social-ecological systems: A critical review, *Ecology and Society,* 11(2): 39. https://www.jstor.org/stable/26266023

Tsouvalis, J. and Waterton, C. (2012) Building 'participation' upon critique: The Loweswater care project, Cumbria, UK, *Environmental Modelling & Software,* 36: 111–121. https://doi.org/10.1016/j.envsoft.2012.01.018

Van Dolah, E.R., Paolisso, M., Sellner, K. and Place, A. (2016) Employing a socio-ecological systems approach to engage harmful algal bloom stakeholders, *Aquatic Ecology,* 50(3): 577–594. https://doi.org/10.1007/s10452-015-9562-z

van Vliet, M., Kok, K. and Veldkamp, T. (2010) Linking stakeholders and modellers in scenario studies: The use of Fuzzy Cognitive Maps as a communication and learning tool, *Futures,* 42(1) : 1–14. https://doi.org/10.1016/j.futures.2009.08.005

Voinov, A. and Bousquet, F. (2010) Modelling with stakeholders, *Environmental Modelling & Software,* 25(11): 1268–1281. https://doi.org/10.1016/j.envsoft.2010.03.007

Voinov, A., Jenni, K., Gray, S.A., Kolagani, N., Glynn, P.D., Bommel, P., Prell, C., Zellner, M., Paolisso, M., Jordan, R., Sterling, E., Schmitt Olabisi, L., Giabbanelli, P.J., Sun, Z., Le Page, C., Elsawah, S., BenDor, T. K., Hubacek, K., Laursen, B.K., Jetter, A. and Smajgl, A. (2018) Tools and methods in participatory modeling: Selecting the right tool for the job, *Environmental Modelling & Software,* 109: 232–255. https://doi.org/10.1016/j.envsoft.2018.08.028

Voinov, A., Kolagani, N., McCall, M.K., Glynn, P.D., Kragt, M.E., Ostermann, F.O., Pierce, S.A. and Ramu, P. (2016) Modelling with stakeholders – Next generation, *Environmental Modelling & Software,* 77: 196–220. https://doi.org/10.1016/j.envsoft.2015.11.016

Wade, E. and Biedenweg, K. (2019) Exploring the diversity of mental models associated with Belize's managed access fisheries policy, *Ocean & Coastal Management*, 178: 104868. https://doi.org/10.1016/j.ocecoaman.2019.104868

Willis, C., Papathanasopoulou, E., Russel, D. and Artioli, Y. (2018) Harmful algal blooms: The impacts on cultural ecosystem services and human well-being in a case study setting, Cornwall, UK, *Marine Policy*, 97: 232–238. https://doi.org/10.1016/j.marpol.2018.06.002

7 Practical considerations for network analysis in research and decision-making

Drawing from diverse approaches for grounded insights

Rebecca Anne Riggs, James Douglas Langston and Mas Davino Sayaza

Introduction

Societies' ability to recognize and understand patterns of connectivity and relationships has proven essential to well-being and innovation. In medicine, finance and engineering, network science offers a way to break-down complex systems and analyse interactions, contributing to knowledge and innovation. In social systems, network analysis can help inform people on behavioural patterns, transfers of knowledge, kinship structures, the spread of ideas and the distributions of power and influence. The emergence of network science can be traced back to the 18th century, with major advances in social sciences in the mid-20th century (Borgatti et al., 2009). Over the years, the use of matrix algebra and graph theory to visualize and analyze network structures and components has become increasingly accessible, supported by the development of network theory. The study of nodes (vertices, points or actors[1]) connected by edges (ties or links) can explain network properties, flows of information and influence and patterns of interactions at different temporal and spatial scales. For those seeking to improve understanding of institutions in natural resource governance, advances in network science have been a welcome addition to interdisciplinary analysis of complex systems. However, practical guidance on the application of network analysis in researching institutions has not kept pace with the anticipated benefits of network approaches. This chapter seeks to address this gap, reflecting on the practical use of network science to uncover institutions in natural resource governance, highlighting complexities with a case study of forest landscape institutions in Indonesia, and discussing future priorities for empirically driven research.

Network analysis and institutions in natural resource governance: what do we know?

The importance of understanding complex systems has long been acknowledged in natural resource governance. The ecosystem concept was central to

DOI: 10.4324/9781032624679-7

early recognition of environmental degradation, building awareness of human–environmental interactions and interest in social–ecological systems. The 1992 Rio Earth Summit put people at the centre of managing natural resources, emphasising the responsibilities of states enacted through partnerships, cooperation and the participation of Indigenous Peoples and Local Communities (IPLCs). In the decades following, studies began to interrogate key actors, roles and responsibilities in natural resource governance, and the ways in which institutions interact with society to change behaviours, mediate conflict and shape environmental change (Leach, Mearns and Scoones, 1999). The concept of polycentric governance systems – that there are multiple centres of decision-making, coordinated but with some degree of autonomy – contributed to more nuanced understandings of how institutions interact across scales in complex systems (Nagendra and Ostrom, 2012; Ostrom, 2010). Network analysis has emerged as useful tool for diving deeper into the actors, rules and distributions of power in decision-making, and the ways in which institutions are formed, coordinate or conflict in natural resource systems.

Network analysis is an umbrella term for various approaches to studying networks. Social network analysis (SNA) and policy network analysis (PNA) are the most common in natural resource governance, with specific theory and methods guiding their implementation. Examples include using SNA to evaluate the influence of formal networks on the formation of informal relationships in collaborative water governance (Huang et al., 2022), understanding the role of institutions on climate adaptive capacity (Jaja, Dawson and Gaudet, 2017) and investigating political structures for policy design using PNA (Brockhaus, Di Gregorio and Carmenta, 2014). These approaches are underpinned by the recognition that social networks influence behaviour (and vice versa), and therefore, the characteristics of networks (actors and interactions) can provide insights into the formal and informal rules that shape behaviours. Actors are often defined by their role and authority; as agents that carry out, contest, influence and are affected by the rules of the game. Interactions can exist across space and time, enabled and constrained by multilayered network configurations (Barnes et al., 2017; Guerrero et al., 2013; McAllister et al., 2020). SNA and PNA are related to actor network theory, which includes non-human agents as actors (Latour, 1996). For example, van der Duim, Ampumuza and Ahebwa (2014) describe gorillas in Ugandan national parks as central actors in an evolving network of externally imposed discourses and ideologies.

A large body of literature guides the use of network analysis in natural resource governance, drawing from theories on social capital, social learning and social processes (Bodin et al., 2016; Crona et al., 2011). These theories provide foundations for considering how institutions are formed and enable influence over values and behaviours (Bodin and Crona, 2009; Huang et al., 2022). Common research areas that might be explored through network analysis and practical steps for research design are discussed below. Broadly, insights on natural resource institutions generated through network analysis can be grouped into three levels of analysis: (1) network configurations; (2) actor positions and (3) community sub-groups. When part of a comprehensive research design, network analysis can provide robust knowledge on how actors interact with each other to form and comply with rules and responsibilities, share information, coordinate, engage in conflict resolution and mobilize resources.

As demonstrated by the case study at the end of this chapter, gathering information on tangible and intangible attributes of ties, natural resource conflicts, demographics, capabilities can also be vital in enriching network insights for institutional analysis.

Network configuration

The structural characteristics of natural resource governance networks can provide important information on the big picture – how actors are situated within governance systems and patterns of social influence. Key features of networks include (1) size – number of nodes in a network; (2) density – how tightly connected nodes are; (3) distance – measures of how easily nodes can connect across the network and (4) patterns of network configuration. Using these concepts, we can capture the number of connected actors with policy influence (Weiss et al., 2012), evaluate redundancy or social capital by calculating network density (Carlsson and Sandström, 2008) and consider how easily ideas and information might diffuse throughout a network (Frank et al., 2023). Network configuration patterns will depend on the research design and objective, discussed below. For example, 'social selection networks' are based on homophily, where actors share identities, values and worldviews. 'Social influence networks' focus on how networks shape and influence actor's perceptions and behaviours (Prell, 2011). Studies interested in social and relational power are likely to focus on the latter, where both formal and informal influence can be explored (Reyhani and Grundmann, 2021).

Actor positions

The position of an actor within a network provides critical information on their capabilities for performing specific functions. Measures related to actor positions can inform which actors perform specific roles, the types of influence that an actor might exercise and the importance of specific actors in carrying out institutional functions. In network analysis, the concept of centrality is one of the most powerful measures for understanding how nodes are situated within a network and the consequent influence on functions and objectives (Freeman, 2002). Four key measures of centrality are (1) degree – how connected a node is to other nodes; (2) betweenness – the ability of a node to connect disconnected nodes; (3) closeness – how easily a node can access other nodes in the network and (4) eigenvector – how close a node is to 'important' nodes (those with high degree). Actors with high degree centrality have been identified as having high authority in decision-making systems, such as village leaders leveraging strong existing social relations as part of their institutional function (Stein, Ernstson and Barron, 2011). Betweenness centrality has been used to identify knowledge brokers and bridging organizations in policy networks, with the capabilities or potential to connect decision-making centres across sectors or scales (Gallemore, Prasti and Moeliono, 2014). Mbaru and Barnes (2017) measured closeness centrality in Kenyan fishing villages to identify that formal leadership and productive assets were important attributes of actors that could spread conservation information efficiently. Calculation of eigenvector centrality scores by Tesfaye et al. (2020) helped to distinguish influential non-government actors in agricultural climate services in Ethiopia, highlighting the influence of intermediaries in institutional functions.

A large body of work confirms the utility of centrality measures in analyzing governance, but they are not without shortcomings. In fragmented networks, centrality may misrepresent a nodes' importance, or overlook nodes that are more optimally situated (example in Figure 7.1). These issues can be overcome by considering 'key players' in a network. Nodes identified as Key Players (Borgatti, 2006) exist due to the way that the network is structured, rather than the node's centrality in the network. To identify these nodes, Borgatti (2006) introduced the key player algorithm, enabling researchers to identify an optimal set of key players for specific objectives – either diffusing something throughout the network or fragmenting a network by removing a node. An

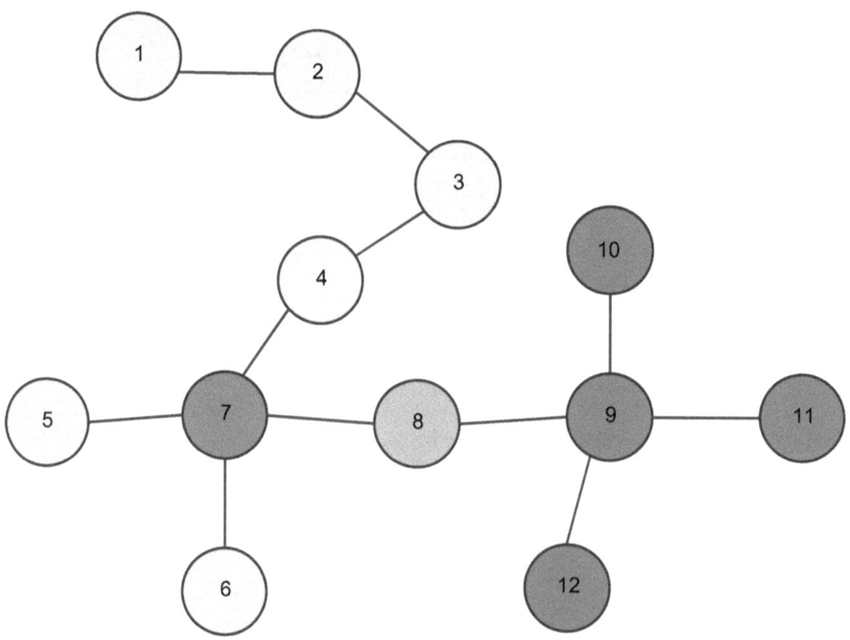

Figure 7.1 Example of key player in a network.

Node 7 is expected to have the highest connectivity, but according to the key player algorithm, Node 8 is more important in connecting all the nodes in the network.

example of key players is provided by a study conducted by Mbaru and Barnes (2017) on rural fishing villages in Kenya. Using the key player algorithm as part of their analysis, Mbaru and Barnes (2017) identify formal leaders as more effective at diffusing conservation-related goals than conventionally targeted experienced fisherman. Another approach is to look at 'brokers' or 'intermediaries' using advanced modelling techniques (Terhorst et al., 2024). Combining these measures in natural resource governance offers deeper insights into how local actors broker new forms of institutions, or how political agents become bricoleurs in institutional reform (Bodin, Crona and Ernstson, 2006; Karambiri et al., 2020).

Community subgroups

Specific measures that focus on communities or sub-groups within a network can provide insights into how different actors dominate or are more peripheral in networks, or the emergence of clusters of informal actors that fill institutional voids. Some of these measures provide information on the entire network; others focus specifically on sub-group structures. The most common measures used to assess clustering are (i) clustering coefficient – the tendency of nodes to cluster together, either as a value for a single node or an average for the network and (ii) modularity – how easily a network could be clustered into communities. A network with high modularity is likely to have stronger connections within clusters and weaker connections between clusters. For example, actors involved in conservation and development initiatives tend to work within sectoral groups creating gaps in institutional arrangements and constraining their influence as documented by Riggs, Langston and Phann (2020) in Northern Cambodia. Another way to evaluate groups in a network is by using (iii) block modelling, a technique that groups nodes based on set or emergent characteristics. Researchers can then observe the types of actors that are grouped into the same or different blocks and characterize ties between blocks. In their global analysis of REDD+ policy networks, Brockhaus and Di Gregorio (2014) showed how dominant relations between key actors can be characterized as conflicting, cooperative or bargaining, depending on the stage of REDD+ policy and implementation. Combining information on group structure with actor positions can be important for understanding the power of weak ties (Granovetter, 1973). Actors may tend to cluster in like-groups, but the actors with ties between disconnected groups may be essential for providing access to information, resources and supporting institutional resilience (Prell, Hubacek and Reed, 2009).

Network analysis in practice

When designing research on institutional analysis that includes some dimension of network analysis, there are several resources that researchers can draw from to carefully consider the purpose of the network analysis, theoretical and practical considerations, and appropriate methodological choice (Guerrero et al., 2020; Prell, 2011). Methodological approaches to network analysis and institutional analysis include a broad spectrum of ontologies, epistemologies, disciplines, scale, research questions, methods and analytical techniques. Researchers may wish to orient themselves in a particular research paradigm to guide their approach. This could be a positivist approach to ground design principles in empirical observations, used within the Bloomington School of institutionalism (Bennett et al., 2018). Studies focused more on how institutions influence power may lean towards constructivist paradigms, or post-structuralism and relational theories more akin with political ecology and critical institutionalism (Bennett et al., 2018). Network analysis is interdisciplinary in nature and can be adopted to fit positivist and constructivist worldviews, or elements of both (i.e., critical realism discussed in Buch-Hansen (2014)). Here, we review empirical guidance on the use of network analysis for natural resource governance, focusing on researching institutions. Research steps are presented linearly, but all stages should be considered in an overall strategy that fits all components together in a coherent design.

Research objective

While research objectives may be broad, for example, to understand interactions between formal and informal institutions or overcome institutional barriers to sustainability, it is useful to consider specific or sub-research objectives that could be met using network analysis. Depending on the research context, objectives that might benefit from network analysis include a general governance diagnostic, preliminary lines of inquiry (i.e., exploratory problem framing) or more precise targeting, focused on an intervention or hypothesis. Network maps can also form boundary objects, created through research processes to facilitate dialogue, enhance understanding of governance systems, and build relationships through co-creation (Zurba, 2022). The key consideration when setting the objective is what type of interaction you are interested in – what are going to be the nodes and ties in the network? Groce et al. (2019) proposed five types of objectives in network analysis for environmental decision-making. We expand on these types in Table 7.1.

Theoretical grounding and making assumptions explicit

Network analysis is primarily concerned with relationships between actors, rather than specific attributes of actors (Borgatti and Ofem, 2010). This distinction is important when combining network insights with institutional analysis. Three assumptions underpin SNA (Knoke and Yang, 2008). First, relations and interactions are often more important in understanding behaviour and performance than actor attributes. Second, networks affect perceptions, values and beliefs of actors through these relations and interactions. Third, relations and interactions are dynamic, with networks continually

Table 7.1 Research objectives in network analysis and examples in researching institutions

Types of research objectives	*Network examples in researching institutions*
1 Identification of key actors or institutions	Key actors and their links, policy networks and coalitions, characterizing interactions between institutions, social processes that influence change
2 Mobilization of resources	Interactions or collaboration that mobilizes or pools resources to overcome institutional barriers, enablers or inhibiters
3 Knowledge flows	Knowledge shared throughout network or by specific actors (bridging organizations and brokers), learning, innovation
4 Processes of social influence	Actors with high influence, types and manifestations of power, position of marginalized actors, factors that influence decision-making or behaviour
5 Processes of cooperation	Type and level of cooperation, collaboration, coordination, conflict, collective action, multi-level governance

changing because of interactions. Some of these assumptions contrast with attribute-based approaches in mainstream institutional analysis, in which performance is attributed to distinguishing characteristics or qualities of institutions (see Chapter 1). Network analysis may therefore be more attuned to critical institutionalism, which emphasizes the dynamic and evolving nature of institutions, shaped by their surroundings (Cleaver, 2012).

Network analysis as a method for evaluating institutions can also result in covering up implicit assumptions that ought to be made explicit. Some of these assumptions are made due to study design (discussed below); others lie in concepts that underpin the research. For example, social capital is a broad and context-dependent concept that can rarely be assigned to one type of relationship. Evaluating weak or strong social capital based on network measures requires clear and explicit definitions (Borg, Toikka and Primmer, 2015). Similarly, when interrogating power dynamics it is useful to be explicit about types or interpretations of power. Theories of power stress that there are multiple interpretations of power (Avelino, 2021). Morrison et al. (2019) distinguish between three types of power in polycentric environmental governance: use of power by *design*, use of *pragmatic* power and use of *framing* power. How power manifests in systems, how it is exercised by actors and how it is entwined with values and knowledge systems can expose how institutions function. Interpretations of power may be different in each of the objectives above (Table 7.1), necessitating careful and explicit theoretical grounding.

Study design

Deciding the parameters of a network study will have direct implications for the analysis and results. Network measures are based on specific variables that need to be systematically documented. These include network boundaries,

Table 7.2 Network analysis parameters

Parameter	*Variable*	*Design considerations*
System	Boundary	Realist (defined by research participants) and nominalist (defined by the researcher). Boundaries may be geographically determined, problem-focused, relationship-based.
Node	Type	Actors, policies, organizations, individuals, occupations, disciplines, species, ecosystem services
	Attributes	Can be assigned to nodes, i.e., governance level, size, sector, mandate, influence, stage of involvement in the network
Ties	Type	Describes the relationship between two nodes. i.e., flows (information, resources), interactions (collaboration, coordination, resistance), relations (kinship), similarities (attributes), power (over, with)
	Directionality	Directed or undirected (reciprocal), i.e., flow of information
	Weight	Numeric or coded value that can be assigned to a tie between two nodes, i.e., strength of cooperation, frequency of interaction

node types, and tie types and attributes (see Table 7.2). Defining the network boundary is a key challenge in network analysis. Networks can either be egocentric (individual – focused on the structure of an actor's immediate network and what that means for that actor) or sociocentric (whole – interested in patterns of interactions between all actors with predetermined boundaries) (Längler, Brouwer and Gruber, 2019). Incomplete data or missing nodes can distort network configurations. While there is no single way to define boundaries, it is widely agreed that boundary decisions must be taken carefully. Freeman (2002) proposed two ways of determining boundaries; realist and nominalist. In the realist approach, network boundaries are defined by research participants, prioritising subjectivity of actors. In the nominalist approach, the researcher defines boundaries based on the research purpose, with clear procedures on which nodes are included and excluded. An alternative approach is deciding on a bounded or unbounded network, the former involving a set number of nodes and the latter an unknown number of nodes and ties. These strategies guide data collection, with some researchers choosing a combination of these approaches.

Additional parameters to decide on are nodes and ties. Nodes may be inclusive of all types of actors (see Riggs, Langston and Phann, 2020) or limited to specific forms or attributes. Studies can gather information on specific or multiple ties within a network and develop criteria for assigning strengths to each tie. Here, researchers face a trade-off between deductive and inductive methodologies. Upfront clarification of the type of tie between nodes can help to build rigorous, replicable networks. Inductive approaches that categorize tie types as they emerge allow for a more complete description of the network but may face difficulties in comparability and replicability. Natural resource governance networks are likely to have several ties that can be distinguished and categorized (see Borgatti and Li, 2009). Each tie represents a distinct network layer but does not exist in isolation. These ties can be analysed in separate networks or in a multi-layer network, discussed in the section on analytical tools.

Data collection methods

Data collection methods in network analysis of institutions can be broadly grouped into six categories: (a) interviews, (b) surveys, (c) document analysis, (d) longitudinal approaches, (e) participatory methods and (f) mixed-methods. Large-scale network analysis using algorithms and digital data (emailing, web-pages, social media) is outside the scope of this chapter. Several factors will influence the researcher's approach, including the questions, scope, resources and current state of knowledge. For all methods, researchers should clarify terms upfront, ensuring consistent responses. Terminology and underpinning concepts like 'actor' and 'relationship' are often difficult to explain in different settings and languages. This can even be hard to explain in familiar settings with different stakeholders. Below, we discuss benefits and limitations of different methods and considerations for network approach.

INTERVIEWS

Interviews typically include researchers asking direct questions to actors situated within the network of interest. For organizations, companies and departments, individuals are asked to represent the actor, and therefore interviews should ideally be held with individuals capable of speaking on behalf of that actor. Specific network questions can be closed or open (see Table 7.3). Open questions require researchers to code and categorize responses and may result in networks

Table 7.3 Open vs closed interview questions

Closed network interview questions	*Open*
– List three organizations that you share information with regarding the community fishery. – How frequently do you share information with these organizations? • At least 1× per month • At least 4× per year • At least 2× per year • At least 1× per year – Is information-sharing two-way? • Yes • No	– Tell us about who you work with at the community fishery – Which actors do you mostly work with? – Can you describe the nature of your relationship with actor x?

consisting of broadly defined ties (i.e., cooperation), or incomplete networks of specific ties. Closed questions are likely to lead to a more complete network but might miss important details about relationships. For example, institutions that sit outside of natural resource governance may have high influence over behaviour and decisions. Interviews allow for more discussion and nuance, but responses to simple network questions can take a long time. This presents a trade-off if combining network analysis with other methods such as power analysis. It may be necessary to prioritize 1–2 ties of interest or limiting responses to lists of 4–5 actors. When conducting interviews, triangulation, saturation and supporting information are key. Researchers may have to make judgements if two actors give opposing responses and be explicit about how these judgements were made.

NETWORK MAP INTERVIEWS

A network map interview visually constructs the network map during the interview, led by the researcher or the interviewee. There is a broad range of approaches to choose from – researchers might provide a template featuring standardized elements (i.e., key actors), interviewees might create a map as a free network drawing, or researchers might ask the interviewee to put themselves at the centre and create an ego-centric network. Network map interviews are similar to participatory methods in that they allow in-depth discussion while visualizing the network, that is, as a boundary object. Interviewees have a tangible object to guide discussion, sharing narratives and 'storied relationships' (Jaspersen and Stein, 2019). With careful and consistent documentation (i.e., assigning tie weights), the network map can then be digitalized for statistical analysis.

QUESTIONNAIRE SURVEYS

Surveys are standardized questions, asking respondents to provide answers in a consistent format. Questions can be a simple and effective way of collecting quantitative data, especially if respondents are spread across different locations. A common technique in network analysis surveys is a name-generator, which asks respondents to free-list actors with whom they share specific ties (participants define relationships). This can be followed by a name interpreter, which asks respondents to identify attributes of the names listed, such as frequency of interaction, relationship length or more descriptive information about the tie (participants qualify relationships). With online survey tools, you can create visual aids to help respondents, and carry answers through to follow-up questions (see Table 7.4). There are specific online survey tools for SNA and software packages that translate and visualize survey results into network maps. These maps can be presented back to participants for verification. Surveys tend to be well suited to homogenous networks and single-purpose studies, where concepts are likely to be clearly understood among respondents and there is less of a need for narrative interpretation of results.

Table 7.4 Survey questionnaire

Survey questionnaire			
1 Select from the actors below, who has influence over how forest resources are used?			
Actor | Level of Influence	*Low*	*Medium*	*High*
Ministry of Forestry			✓
Customary leaders		✓	
First Nations Forestry Council	✓		
Religious leaders			✓
Community Forest Association		✓	
1 Name actors that set rules on using forest resources (name generator)			
a How closely do you obey these rules? (name interpreter)			
b Choosing from the options below, characterize your relationship with the actors listed:			
Actor|Level of Influence	*Information sharing*	*Collaboration*	*Coordination*
Ministry of Forestry	✓		✓
Company X	✓	✓	
First Nations Forestry Council			
Industry Association			
Community Forestry Association	✓	✓	

DOCUMENT ANALYSIS

Networks created from secondary sources (policy documents, archival records, literature, media) can lead to a more extensive representation of formal governance networks, avoiding issues of bias and high cost in interviews and surveys. While document analysis is more common in a mixed methods approach, the review of policies and supporting documents can be sufficient to identify key actors and their interaction over time, participation in policy processes and policy cohesion (Yu et al., 2022). Network data can also be constructed from organizational datasets (i.e. conservation or development projects) showing ties at the national and/or stakeholder level, and the distribution of power geographically (Shin et al., 2022). Standard document analysis methods apply to network analysis; on deciding the initial sources, researchers can systematically code qualitative content to identify variables used in the analysis (nodes and ties), guided by inductive or deductive approaches. Coded data can then be quantified, such as assigning weights to ties. Document analysis is well suited to policy and formal institutional arrangements but is likely to miss informal relationships and oral transmission of knowledge – which may be more common in IPLCs.

LONGITUDINAL APPROACHES

Longitudinal network analysis involves examining changes in network structure over time, providing insights to how actors change their ties in a network in response or in anticipation to events and objectives. The preferred data collection method for longitudinal analysis is repeating the same method over time, such as interviews conducted annually with a panel or surveys distributed to specific actors following specific events or over a time-series, referred to as waves. Other options are to extract data from media articles or archival data over specific periods. With two or more points in time, longitudinal datasets can be simulated using various models, analyzing network dynamics. For example, Aung and Lim (2021) used longitudinal data collected through repeated surveys to show organizational responses to flood disasters in Myanmar. Longitudinal studies are high cost and often have participant retention challenges, making it easier to use simulations or models to explore network changes over time.

PARTICIPATORY METHODS

Participatory network analysis aims to elicit these benefits while co-developing a visual network map and engaging participants in a discussion about the network produced. Different participatory network approaches have emerged over the years. The Net-Map method, developed by Schiffer and Hauck (2010) has been used in different natural resource governance contexts, intended as a low-cost low-tech tool for workshops or group settings. Net-Map follows a process of identifying key stakeholders on cards, drawing and labelling ties, building influence towers, documenting goals, followed by participatory analysis of the network built (https://netmap.wordpress.com). The advantage of participatory network analysis is the co-learning that takes place during the process. Discussion allows assumptions to be challenged and made explicit, and participants can interrogate terms, ask questions and take ownership of the progress. In their experiences of using Net-Map to support the implementation of Payments for Ecosystem Services, Schröter et al. (2018) reflect on benefits in creating common understanding, the value of participant creativity and trust, and the limitations of time and resources needed. However, sensitivities due to power and cultural relations can also mean participatory methods are not always conducive to institutional analysis (Bisaro, Roggero and Villamayor-Tomas, 2018). In studies where participants are geographically dispersed or facilitators may have difficulty in respect to cultural and power differences, other methods may be more appropriate.

Mixed-methods

When researching institutions, a mixed-methods approach is helpful for developing an understanding of the nature of relationships, experiences and contextual factors that affect networks. A common approach to mixed-method research design is to begin by collecting qualitative data to build a rich understanding of the research question and develop a hypothesis, and then collect quantitative data using a questionnaire. This resembles retroductive or abductive reasoning, which is based on iteration between inductive empirical observation and deductive testing. In reverse, interviews can follow surveys, enhancing interpretation of results. These retroductive approaches are ideal for generating more robust understandings of system 'causality' (Buch-Hansen, 2014). Other more creative approaches include: combining PPGIS (Public Participatory Geographic Information Systems) and questionnaires to map relationships between stakeholder and social-cultural values of ecosystem services (Barraclough, Cusens and Måren, 2022); comparing de jure networks according to policy documents with de facto networks described in interviews (Sari et al., 2019); using household demographic surveys to identify characteristics of key players (Mbaru and Barnes, 2017); combining participatory workshops with priority and value questionnaires, documented by Buckingham et al. (2021) in a case study of restoration governance in Rwanda.

Analytical tools

Descriptive analysis of networks can be simple and useful – there is a lot to be gained from understanding who is involved in a network, how they are connected, and the extent and structure of institutional arrangements. However, there are numerous tools and techniques that allow for a more rigorous and complex analysis, using the range of network measures available, building from existing studies and contributing to a growing evidence-base. Entry-level programmes include open-source software Gephi (https://gephi.org) and Cytoscape (http://www.cytoscape.org), Kumi (https://www.kumu.io) and UCINET (http://www.analytictech.com/archive/ucinet.htm), which have free trials available; and several programming languages (R, Python) have network packages for more advanced analysis. Standard network measures can be calculated using most network analysis platforms.

Advanced analytical methods offer statistical rigour for investigating interactions and causality, but require careful attention to assumptions in the dataset and models (Guerrero et al., 2020). These include: exponential random graph models (ERGMs) to investigate randomness or interdependence of ties; stochastic actor-oriented model (SOAMs) to explore the co-evolution of network structures and actor behaviours (referred to as longitudinal network analysis); autocorrelation models to predict node characteristics based on ties with other nodes with the same characteristic (de Nooy, 2013); and agent-based modelling (ABM) for analysing causal relationships, behaviours and decisions of governance actors (Bourceret, Amblard and Mathias, 2021). In studies of institutional arrangements, SOAMs have been used to demonstrate how close relationships and mutual trust motivate cooperation (Aung and Lim, 2021). ERGMs can help to analyze multi-layered ties in a network, eliciting how ties might be nested and influence each other (McAllister et al., 2020). Advanced approaches enable researchers to capture dynamism, create scenarios (ABM) and test hypotheses about causality (Bodin et al., 2019).

Reporting results and ethical considerations

The power of network analysis often lies in the tools that enable visualization. Sociograms (graphic representations of networks) can be constructed to convey key results, including node size, colour, structure and patterns of group interactions. There is also a risk of visual manipulation, which can overstate or skew results. Missing or flawed data, imperfect measurements, missing links can all hinder the accuracy of network studies on institutions, misleading interpretation of results. Reporting on each of the steps above can help to minimize inaccuracies, providing clear descriptions on data collection and analysis, including which measurements were used in the study and why. Each of the data collection techniques above come with their own set of considerations for ethical practice. Who has the authority to speak on behalf of whom or represent views? How might results be misinterpreted or misused? Increasing visibility of informal governance systems may have undesirable effects, participatory actions to enable social change may lead to unmet expectations. Ethical considerations may include whether/how to identify actors and their ties, triangulate opposing perspectives, or privilege marginalized voices. Researching institutions for the purpose of public disclosure of research findings risks selective information sharing which can distort results. Integrating analysis with other methods can help to address these shortfalls, but it is important to acknowledge caveats and limitations of network studies when reporting results.

Case study: forest landscape governance in Maluku, Indonesia

In Maluku province in Eastern Indonesia, forests are valued for their contributions to biodiversity conservation, ecosystem services, agroforestry and timber incomes, and household sustenance. Decisions over how to sustain these values are made by a diverse set of actors at different levels: central government agencies in Jakarta, provincial forestry departments, Manusela National Park, local communities and the involvement of many other public, private and non-government actors. Forests are governed through both customary (*Adat*) and formal state institutions, with varying degrees of harmony and conflict over the rules of the game. Customary actors include *Adat* leaders with various positions in village governance, with roles such as regulating forest and land ownership, and village council members. As part of a long-term transdisciplinary research programme, we conducted network analysis of forest landscape governance in North Seram in June 2023. We sought to understand how state and customary actors interacted in decisions over forests, and what potential role the government-led Social Forestry programme could play in bridging these actors. The fuzzy boundaries between state and nonstate forest areas and mosaic land-use patterns in North Seram make the concept of a landscape an appropriate entry point for understanding how state and customary institutions interact in forest governance. Here, forest landscape governance describes the polycentric decision-making systems affecting a landscape where forests are a central feature.

We collected data through semi-structured interviews with actors with a known or expected role in forest governance. Several years of qualitative research and collaboration helped to set boundaries, identify interviewees and refine the objective of the network analysis. For example, we interviewed both religious leaders and fisheries actors that do not have explicit roles in forest governance but were identified as important in the broader social–ecological system. Interviewees were asked to name which actors they work with to perform their roles relating to forests and describe the nature of the work (cooperation network with undirected ties). We also asked which actors were perceived to have the most influence over forests, and for interviewees to rank which institutions guide their decision-making (state, customary, religion, other). Analysis of 21 interviews in Kumi provided key network measures and visualization of the cooperation network.

The network visualization (Figure 7.2) is shows nodes coloured by governance level. The four actors that scored high on centrality measures (degree, betweenness and closeness) were the provincial government agencies responsible for forestry, conservation and social forestry, as well as the village government. Sitting just outside of this core network, with high eigenvector scores but lower scores for other centrality measures, were the customary actors and social forestry groups (visible as labelled nodes in Figure 7.2). There was no notable difference between institutions that guided interviewee

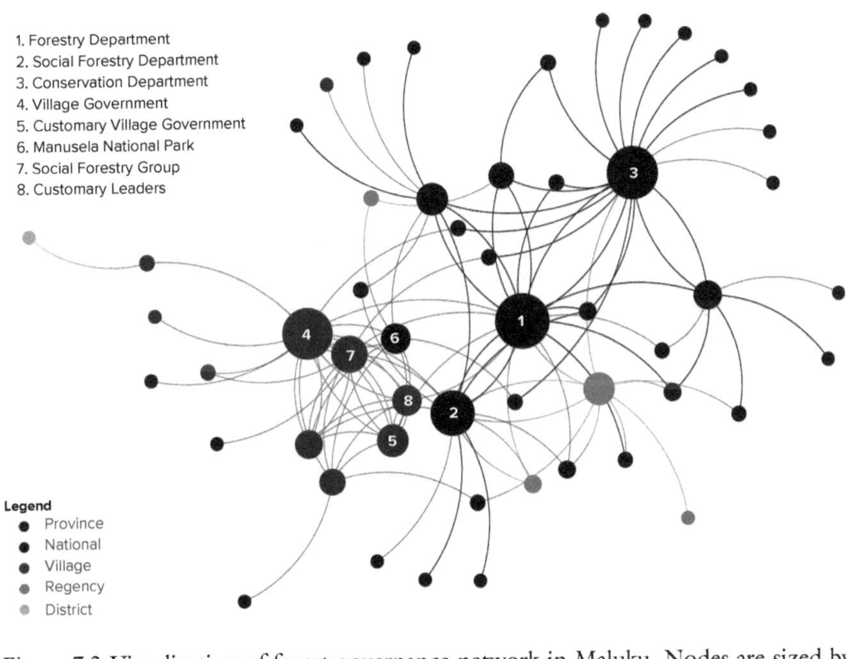

Figure 7.2 Visualization of forest governance network in Maluku. Nodes are sized by degree centrality and coloured by governance level.

Source: Authors.

decisions – state, customary, community and the church were all reported to be important when actors made decisions. The actor with the highest perceived influence over natural resources was the village government. For most interviewees, this included both customary and formal leaders as a combined entity, whose role, responsibilities and capabilities were to make decisions for the community across all aspects of life. When separated by interviewee type, there was a clear difference in perceptions of which actor was considered the most influential. Government agencies were more likely to choose the Provincial Government or Ministry of Environment and Forestry, and non-government actors at the village level were more likely to choose village government. These results, combined with the centrality measures and sociogram, show a disconnect between state and village-level institutions, and highlight the potential for Social Forestry to provide a bridge. The centrality of the conservation and forestry agencies suggest that to be effective, the Social Forestry agency must leverage these agencies, expanding their reach and harnessing existing resources.

Reflections on the practical application of network analysis

In this case study, the practicalities of using network analysis in researching forest institutions in Maluku, Indonesia, were complicated. Researchers were familiar with context and engaged local partners. Yet the complexity of interconnected institutions posed a challenge for data collection and analysis. Interview questions were simplified and adapted to each participant, and results were analysed to enrich and inform years of ethnographic and participatory research. In this way, the network analysis was a boundary object – provoking reaction among collaborative, transdisciplinary researchers from diverse disciplines.

During interviews, we learned that several small agencies work on short-term projects, making it difficult for village-level actors to clearly identify actors with whom they cooperate. The remoteness of the study site and distance from the provincial capital means interactions are infrequent and connections are weak or based on past experiences. This posed a challenge in both data collection and analysis – what should count as a tie in this network? The legal system of village governance in Maluku attempts to combine both customary and state institutions, which can lead to ambiguity over the roles and responsibilities of actors. At times, it was difficult for participants to distinguish between the formal and informal governance arrangements. Researchers had to choose whether to disrupt the flow of conversation for clarification or attempt to interpret ambiguous terms ex-post.

Despite these limitations, the results of the network analysis, combined with qualitative interview data, provide important insights that could help to guide the Social Forestry programme and understand the rules of the game. We learnt that structures set up to guide social forestry implementation were not yet functioning, and participants welcomed the opportunity for research to help raise the profile of the programme and stimulate progress. We found that leveraging the external connections to companies and NGOs held by provincial government agencies could be vital to closing gaps in resources and capacities in forest management. The lack of connectivity between provincial agencies and the National Park (sharing village boundaries) suggests a missed opportunity to developing more durable and frequent interactions. The closely entwined village-level institutions demonstrate the need to recognize the polycentric decision-making structures within the landscape, while also respecting the authority of the village government. Finally, the results showing different perceptions of who holds authority over forest-related decisions, highlighting that issues of justice, tenure and rights must be grappled with for long-term coherent natural resource governance.

Conclusion

Network analysis can serve many roles for those interested in researching institutions – from a simple visual depiction of key actors to a predictive model of decision-making and influence. In general, fundamental lessons emerge: select methods according to context and objectives, complementary data is helpful, and understanding limitations is important for interpreting results. The case study of Maluku forest landscape governance highlights the value of considering informal and socially embedded institutions in network analysis. It also re-enforces the need to ground studies in context, and the misrepresentations that might emerge without multi-level, qualitative research design. The power of network analysis to generate new insights or deepen engagement and dialogue will depend on the research objective and design. A solid theoretical foundation that covers network theory, power and institutions can help to ensure researchers maximize the contributions of empirical studies.

Researchers are fortunate to have a large body of interdisciplinary work to draw from when designing network studies – the references in this chapter intend to serve as a starting point. Yet implementing these designs often requires muddling through, especially when engaging with diverse actors with different capabilities and power. How questions are perceived and answered can vary greatly, and there are risks involved in making implicit or hidden relationships explicit. Further research that helps to shed light on navigating these challenges could help to develop robust and comparable studies. For example, how do mental models and cognitive biases influence data collection methods such as name-generators and free-listing? How do we capture the influence of invisible or asymmetric ties in a network? These questions are beginning to be asked in diverse applications of network analysis (Brass, 2022). Researching institutions provides an opportunity to ground these questions in empirical understanding of natural resource governance, advancing theoretical approaches, analytical techniques and practice-based learning.

Note

1 In this chapter, the term actor is used to describe an individual or entity (i.e., stakeholder, community).

References

Aung, T.M. and Lim, S. (2021) Evolution of collaborative governance in the 2015, 2016, and 2018 Myanmar flood disaster responses: A longitudinal approach to a network analysis, *International Journal of Disaster Risk Science*, 12(2): 267–280. https://doi.org/10.1007/s13753-021-00332-y

Avelino, F. (2021) Theories of power and social change. Power contestations and their implications for research on social change and innovation, *Journal of Political Power*, 14(3): 425–448. https://doi.org/10.1080/2158379X.2021.1875307

Barnes, M.L., Bodin, Ö., Guerrero, A.M., McAllister, R.R.J., Alexander, S.M. and Robins, G (2017) The social structural foundations of adaptation and transformation in social–ecological systems, *Ecology and Society*, 22(4): 16. Available at https://www.jstor.org/stable/26798997

Barraclough, A.D., Cusens, J. and Måren, I. E (2022) Mapping stakeholder networks for the co-production of multiple ecosystem services: A novel mixed-methods approach, *Ecosystem Services*, 56: 101461. https://doi.org/10.1016/j.ecoser.2022.101461

Bennett, A., Acton, L., Epstein, G., Gruby, R.L. and Nenadovic, M. (2018) Embracing conceptual diversity to integrate power and institutional analysis: Introducing a relational typology, *International Journal of the Commons*, 12(2): 330–357. https://doi.org/10.18352/ijc.819

Bisaro, A., Roggero, M. and Villamayor-Tomas, S. (2018) Institutional analysis in climate change adaptation research: A systematic literature review, *Ecological Economics*, 151: 34–43. https://doi.org/10.1016/j.ecolecon.2018.04.016

Bodin, Ö., Alexander, S.M., Baggio, J., Barnes, M.L., Berardo, R., Cumming, G.S., Dee, L.E., Fischer, A.P., Fischer, M., Mancilla Garcia, M., Guerrero, A.M., Hileman, J., Ingold, K., Matous, P., Morrison, T.H., Nohrstedt, D., Pittman, J., Robins, G. and Sayles, J. S. (2019) Improving network approaches to the study of complex social–ecological interdependencies, *Nature Sustainability*, 2(7): 551–559. https://doi.org/10.1038/s41893-019-0308-0

Bodin, Ö., Crona, B. and Ernstson, H. (2006) Social networks in natural resource management: What is there to learn from a structural perspective?, *Ecology and Society*, 11(2): r2. https://www.ecologyandsociety.org/vol11/iss2/resp2/

Bodin, Ö., and Crona, B.I. (2009) The role of social networks in natural resource governance: What relational patterns make a difference?, *Global Environmental Change*, 19(3): 366–374. https://doi.org/10.1016/j.gloenvcha.2009.05.002

Bodin, Ö., Robins, G., McAllister, R.R., Guerrero, A.M., Crona, B., Tengö, M. and Lubell, M. (2016) Theorizing benefits and constraints in collaborative environmental governance: A transdisciplinary social-ecological network approach for empirical investigations, *Ecology and Society*, 21(1): 40. https://doi.org/10.5751/ES-08368-210140

Borg, R., Toikka, A. and Primmer, E. (2015) Social capital and governance: A social network analysis of forest biodiversity collaboration in Central Finland, *Forest Policy and Economics*, 50: 90–97. https://doi.org/10.1016/j.forpol.2014.06.008

Borgatti, S.P. (2006) Identifying sets of key players in a social network, *Computational & Mathematical Organization Theory*, 12(1): 21–34. https://doi.org/10.1007/s10588-006-7084-x

Borgatti, S.P. and Li, X. (2009) On social network analysis in a supply chain context, *Journal of Supply Chain Management*, 45(2): 5–22. https://doi.org/10.1111/j.1745-493X.2009.03166.x

Borgatti, S.P., Mehra, A., Brass, D.J. and Labianca, G. (2009) Network analysis in the social sciences, *Science*, 323(5916): 892–895. https://doi.org/10.1126/science.1165821

Borgatti, S.P. and Ofem, B. (2010) Social network theory and analysis, in Daly, A. (ed.) *Social Network Theory and Educational Change*, Cambridge, MA: Harvard Education Press, pp. 17–29.

Bourceret, A., Amblard, L. and Mathias, J.-D. (2021) Governance in social-ecological agent-based models: A review, *Ecology and Society*, 26(2): 38. https://doi.org/10.5751/ES-12440-260238

Brass, D.J. (2022) New developments in social network analysis, *Annual Review of Organizational Psychology and Organizational Behavior*, 9(1): 225–246. https://doi.org/10.1146/annurev-orgpsych-012420-090628

Brockhaus, M., and Di Gregorio, M. (2014) National REDD+ policy networks from cooperation to conflict, *Ecology and Society*, 19(4): 14. https://www.jstor.org/stable/26269660

Brockhaus, M., Di Gregorio, M. and Carmenta, R. (2014) REDD+ policy networks exploring actors and power structures in an emerging policy domain, *Ecology and Society*, 19(4): 29. Available at https://www.jstor.org/stable/26269649

Buch-Hansen, H. (2014) Social network analysis and critical realism, *Journal for the Theory of Social Behaviour*, 44(3):306–325. https://doi.org/10.1111/jtsb.12044

Buckingham, K., Arakwiye, B., Ray, S., Maneerattana, O. and Anderson, W. (2021) Cultivating networks and mapping social landscapes: How to understand restoration governance in Rwanda, *Land Use Policy*, 104: 104546. https://doi.org/10.1016/j.landusepol.2020.104546

Carlsson, L. and Sandström, A. (2008) Network governance of the commons, *International Journal of the Commons*, 2(1): 33–54. https://doi.org/10.18352/ijc.20

Cleaver, F. (2012) *Development through Bricolage: Rethinking Institutions for Natural Resource Management*, London: Earthscan. https://doi.org/10.4324/9781315094915

Crona, B., Ernstson, H., Prell, C., Reed, M. and Hubacek, K. (2011) Combining social network approaches with social theories to improve understanding of natural resource governance, in Bodin, O. and Prell, C. (eds.) *Social Networks and Natural Resource Management: Uncovering the Social Fabric of Environmental Governance*, Cambridge: Cambridge University Press, pp. 44–71.

de Nooy, W. (2013) Communication in natural resource management agreement between and disagreement within stakeholder groups, *Ecology and Society*, 18(2): 44. https://www.jstor.org/stable/26269321

Frank, K.A., Chen, T., Brown, E., Larsen, A. and Baule, W.B.J. (2023) A network intervention for natural resource management in the context of climate change, *Social Networks*, 75: 55–64. https://doi.org/10.1016/j.socnet.2022.03.003

Freeman, L.C. (2002) Centrality in social networks: Conceptual clarification, in Scott, J. (ed.) *Social Networks: Critical Concepts in Sociology*, London: Routledge, pp. 238–264.

Gallemore, C.T., Prasti, H.R.D., and Moeliono, M. (2014) Discursive barriers and cross-scale forest governance in Central Kalimantan, Indonesia, *Ecology and Society*, 19(2): 18. https://doi.org/10.5751/ES-06418-190218

Granovetter, M.S. (1973) The strength of weak ties, *American Journal of Sociology*, 78(6): 1360–1380. https://doi.org/10.1086/225469

Groce, J.E., Farrelly, M.A., Jorgensen, B.S. and Cook, C.N. (2019) Using social-network research to improve outcomes in natural resource management, *Conservation Biology*, 33(1): 53–65. https://doi.org/10.1111/cobi.13127

Guerrero, A.M., Barnes, M., Bodin, Ö., Chadès, I., Davis, K.J., Iftekhar, M.S., Morgans, C. and Wilson, K.A. (2020) Key considerations and challenges in the application of social-network research for environmental decision making, *Conservation Biology*, 34(3): 733–742. https://doi.org/10.1111/cobi.13461

Guerrero, A.M., McAllister, R.R.J., Corcoran, J. and Wilson, K.A. (2013) Scale mismatches, conservation planning, and the value of social-network analyses, *Conservation Biology*, 27(1): 35–44. https://doi.org/10.1111/j.1523-1739.2012.01964.x

Huang, C., Yi, H., Chen, T., Xu, X. and Chen, S. (2022) Networked environmental governance: Formal and informal collaborative networks in local China, *Policy Studies*, 43(3): 403–421. https://doi.org/10.1080/01442872.2020.1758306

Jaja, J., Dawson, J. and Gaudet, J. (2017) Using social network analysis to examine the role that institutional integration plays in community-based adaptive capacity to climate change in Caribbean small island communities, *Local Environment*, 22(4): 424–442. https://doi.org/10.1080/13549839.2016.1213711

Jaspersen, L.J., and Stein, C. (2019) Beyond the matrix: Visual methods for qualitative network research, *British Journal of Management*, 30(3): 748–763. https://doi.org/10.1111/1467-8551.12339

Karambiri, M., Brockhaus, M., Sehring, J. and Degrande, A. (2020) 'We are not bad people' – Bricolage and the rise of community forest institutions in Burkina Faso, *International Journal of the Commons*, 14(1): 525–538. https://doi.org/10.5334/ijc.1061

Knoke, D. and Yang, S. (2008) *Social Network Analysis*, Thousand Oaks, CA: SAGE.

Längler, M., Brouwer, J. and Gruber, H. (2019) Data collection for mixed method approaches in social network analysis, in Froehlich, D.E., Rehm, M. and Rienties, B.C. (eds.) *Mixed Methods Social Network Analysis*, London: Routledge, pp. 25–37. https://doi.org/10.4324/9780429056826-4

Latour, B. (1996) On actor-network theory: A few clarifications, *Soziale Welt*, 47: 369–381.

Leach, M., Mearns, R. and Scoones, I. (1999) Environmental entitlements: Dynamics and institutions in community-based natural resource management, *World Development*, 27(2): 225–247. https://doi.org/10.1016/S0305-750X(98)00141-7

Mbaru, E.K. and Barnes, M.L. (2017) Key players in conservation diffusion: Using social network analysis to identify critical injection points, *Biological Conservation*, 210: 222–232. https://doi.org/10.1016/j.biocon.2017.03.031

McAllister, R.R.J., Kruger, H., Stenekes, N. and Garrard, R. (2020) Multilevel stakeholder networks for Australian marine biosecurity: Well-structured for top-down information provision, requires better two-way communication, *Ecology and Society*, 25(3): 18. https://doi.org/10.5751/ES-11583-250318

Morrison, T., Adger, W.N., Brown, K., Lemos, M.C., Huitema, D., Phelps, J., Evans, L., Cohen, P., Song, A. and Turner, R. (2019) The black box of power in polycentric environmental governance, *Global Environmental Change*, 57: 101934. https://doi.org/10.1016/j.gloenvcha.2019.101934

Nagendra, H. and Ostrom, E. (2012) Polycentric governance of multifunctional forested landscapes, *International Journal of the Commons*, 6(2): 104–133. https://doi.org/10.18352/ijc.321

Ostrom, E. (2010) Beyond markets and states: Polycentric governance of complex economic systems, *The American Economic Review*, 100(3): 641–672. https://doi.org/10.1257/aer.100.3.641

Prell, C. (2011) *Social Network Analysis: History, Theory and Methodology*, London: Sage Publications.

Prell, C., Hubacek, K. and Reed, M. (2009) Stakeholder analysis and social network analysis in natural resource management, *Society and Natural Resources*, 22(6): 501–518. https://doi.org/10.1080/08941920802199202

Reyhani, M.N. and Grundmann, P. (2021) Who influences whom and how in river-basin governance? A participatory stakeholder and social network analysis in Zayandeh-Rud basin, Iran, *Environmental Development*, 40: 100677. https://doi.org/10.1016/j.envdev.2021.100677

Riggs, R.A., Langston, J.D. and Phann, S. (2020) Actor network analysis to leverage improvements in conservation and development outcomes in Cambodia, *Ecology and Society*, 25(4): 28. https://doi.org/10.5751/ES-11854-250428

Sari, D.A., Sayer, J., Margules, C. and Boedhihartono, A.K. (2019) Determining the effectiveness of forest landscape governance: A case study from the Sendang landscape, South Sumatra, *Forest Policy and Economics*, 102: 17–28. https://doi.org/10.1016/j.forpol.2019.01.014

Schiffer, E. and Hauck, J. (2010) Net-Map: Collecting social network data and facilitating network learning through participatory influence network mapping, *Field Methods*, 22(3): 231–249. https://doi.org/10.1177/1525822X10374798

Schröter, B., Hauck, J., Hackenberg, I. and Matzdorf, B. (2018) Bringing transparency into the process: Social network analysis as a tool to support the participatory design and implementation process of payments for ecosystem services, *Ecosystem Services*, 34: 206–217. https://doi.org/10.1016/j.ecoser.2018.03.007

Shin, S., Park, M.S., Lee, H. and Baral, H. (2022) The structure and pattern of global partnerships in the REDD+ mechanism, *Forest Policy and Economics*, 135: 102640. https://doi.org/10.1016/j.forpol.2021.102640

Stein, C., Ernstson, H. and Barron, J. (2011) A social network approach to analyzing water governance: The case of the Mkindo catchment, Tanzania, *Physics and Chemistry of the Earth, Parts A/B/C, 36*(14): 1085–1092. https://doi.org/10.1016/j.pce.2011.07.083

Terhorst, A., Wang, P., Lusher, D., Bolton, D. and Elsum, I. (2024) Broker roles in open innovation, *Journal of Open Innovation: Technology, Market, and Complexity*, 10(1): 100186. https://doi.org/10.1016/j.joitmc.2023.100186

Tesfaye, A., Hansen, J., Radeny, M., Belay, S. and Solomon, D. (2020) Actor roles and networks in agricultural climate services in Ethiopia: A social network analysis, *Climate and Development*, 12(8): 769–780. https://doi.org/10.1080/17565529.2019.1691485

van der Duim, R., Ampumuza, C. and Ahebwa, W.M. (2014) Gorilla tourism in Bwindi impenetrable National Park, Uganda: An actor-network perspective, *Society & Natural Resources*, 27(6): 588–601. https://doi.org/10.1080/08941920.2014.901459

Weiss, K., Hamann, M., Kinney, M. and Marsh, H. (2012) Knowledge exchange and policy influence in a marine resource governance network, *Global Environmental Change*, 22(1): 178–188. https://doi.org/10.1016/j.gloenvcha.2011.09.007

Yu, H., Wang, Y., Yin, W. and Li, D. (2022) Policy network analysis of China's ocean ranching policy: Network structure, actors and interaction, *Marine Policy*, 140: 105070. https://doi.org/10.1016/j.marpol.2022.105070

Zurba, M. (2022) Boundary work as a concept and practice in human geography, *Journal of Cultural Geography*, 39(1): 1–7. https://doi.org/10.1080/08873631.2021.2011684

8 Deconstructing local institutions in a forest landscape

An ethnographic case study from India

Sumana Datta

Introduction

In academic research, ethnography is a popular approach employed by scholars within the realm of natural resource governance to investigate and understand institutional frameworks, including but not limited to land (Li, 2014), conservation (Brockington, 2002) and water (Baviskar, 1996). Over the course of a century, ethnography has undergone evolution and has been practised in various iterations by scholars beyond conventional anthropology (Atkinson and Hammersley, 2007). However, at its core, it remains anchored in a process wherein the researcher closely watches 'what happens, listens to what is said, and/or asks questions through informal and formal interviews over the course of a year or more' (Atkinson and Hammersley, 2007: 3) in 'naturally occurring settings' (Brewer, 2000: 6). Thus, ethnography entails a 'methodological commitment to spend long periods in one localised setting' (Gupta and Ferguson, 1997: 4).

Ethnography typically adopts an unstructured approach to research, including research design, data collection methods and data interpretation. The researcher embarks on the research with a specific interest but maintains a flexible approach, a concept termed a 'foreshadowed problem' by Malinowski (Atkinson and Hammersley, 2007). Whilst participant observation is the most intensely used in ethnography, researchers often complement it with supplementary qualitative tools such as semi-structured ethnographic interviews, focus group discussions and document analysis alongside quantitative methods. Furthermore, categories used for interpretation emerge from the iterative process of data analysis.

In this chapter, I reflect on the process of employing an ethnographic method to untangle the intricacies of the institutional framework for forest governance within two villages in the West Bengal state of India, represented as Patrasayer and Sonamukhi.[1] It is pertinent to note that I present a 'reflective account' from the perspective of a 'native' female ethnographer (Kraldy, 2002) as an Indian student attending a university situated in the global north and conducting field research in the global south.

My objective was to examine the effectiveness of the institution of joint forest management in India, which was introduced through an executive

DOI: 10.4324/9781032624679-8

order issued by the Ministry of Forest and Environment[2] in 1990. Joint forest management embodies a co-management model of decentralised governance wherein state agencies, notably the Forest Department, forge partnerships with local communities, establishing 'single purpose user committees' (Manor, 2004: 184) to engage them in protection and management endeavours. This institutional framework is a formal arrangement structured into multiple layers, ranging from local-level forest protection committees to the hierarchies of the Forest Department. My research focused on village-level forest protection committees with dual objectives: first, to evaluate their effectiveness in fostering inclusive local forest governance by ensuring procedural fairness to decision-making processes and allocating equitable benefits derived from forest protection efforts, including access to non-timber forest products for livelihoods and cash returns from timber harvesting and wage works; second, to investigate the functional devolution of power to local forest protection committees and their integration with other layers of governance.

My field research extended over a period of 13 months, commencing in January 2010 and concluding in February 2011. Within the ethnographic approach, I incorporated household surveys, with three repeat round surveys, to gather quantitative descriptions of household forest reliance. The quantitative method aimed to investigate the economic dynamics of forest interaction and assess equity in access to benefits derived from forest resources.

Why ethnography?

Studying an institution like a forest management committee presents challenges. On the one hand, it is deeply enmeshed within its native socio-cultural and political structures, being reproduced, regularised and resisted through the everyday practices and processes of social life (Cleaver, 2012). On the other, it is part of a larger system, namely the Forest Department, where decision-making occurs at multiple layers. Thus, structurally, it functions as both a bureaucratic entity and a socially embedded institution, contrasting with the bureaucratic institution concept (Cleaver, 2002) discussed in Chapter 1. My research primarily focused on the functioning of the forest protection committee and examined how its decision-making processes impacted and were influenced by other layers of decision-makers within the Forest Department.

Ethnography emerges as the most suitable research approach for unravelling these complex and constantly evolving institutions, such as power relations and kinship, which may not appear to be directly connected with the institution of forest governance (Sandström, 2009). Moreover, an integral aspect of comprehending the institution's operation lies in grasping the nuances of distributive justice and procedural fairness within the decision-making processes. Whilst distributive justice seeks fair allocation of cost, risks and benefits amongst different groups, procedural fairness desires representation of stakeholders in the decision-making. Together, both forms of justice intend to safeguard the rights and interests of marginalised groups, including women, landless families and ethnic minorities (McDermott, Mahanty and Schreckenberg, 2013). In the community forestry programmes in Nepal, for example, the poor bore the maximum costs of forest regulation due to their inability to influence decision-making (Sunam and McCarthy, 2010).

An ethnographic approach gathers data on everyday functioning, allowing the researcher to immerse themselves in witnessing the decision-making procedures of institutions within a setting that closely mirrors natural conditions (Atkinson and Hammersley, 2007). I qualify 'closely' as the researcher's presence may initially influence participants' behaviours, albeit this effect tends to minimise over time.

Accessing 'subjects'

> The practical constraints often limit the choices or create favourable opportunities for ethnographic research.
>
> (Gupta and Ferguson, 1997: 4)

I identified both research villages through pilot visits to ten villages in the northern Bankura region. This occurred during a period when the majority of the South West Bengal region, with the exception of a small area along the northern border, was heavily affected by the leftist Naxalite movement. Officials whom I consulted for formal research permission at the state and district headquarters advised me to select an area from the northern part of the Bankura district due to safety considerations. During my pilot visit, I was introduced to the village by the Forest Guard, the frontline representative of the state Forest Department, serving as the crucial intermediary between the local communities and the government. In the region where state surveillance and the influence of groups like Naxal (with extreme left ideology) loom large, gaining access to the village required more than just personal identification. His introduction not only authenticated my identity but also legitimised the objectives of my research visit in the eyes of the villagers. Thus, by vouching for me, the Forest Guard helped alleviate any apprehensions the villagers might have had about my presence in the area. However, this positioned me in the eyes of the villagers as someone with a connection to both the Forest Department and the influential power centres within the village. Subsequently, this portrayal created a sense of distance between me and the common villagers. Later, as I began interacting with the people and articulating the objectives of my research, I was able to bridge the gap created by the initial impression. Thus, active engagement with the community helped to transcend the initial assumptions and establish meaningful connections grounded in mutual respect and shared objectives.

Both villages of my study comprised a small population, each with fewer than 50 households distributed across multiple hamlets. In the heterogeneous village Sonamukhi, hamlet formation was predominantly based on caste criteria. However, the determinants for hamlet formation amongst tribal communities in both villages eluded clarity initially, yet subsequent observations revealed discernible socio-economic disparities amongst these hamlets.

In Patrasayer, I was first introduced to the village representative, who was elected to the village council (locally referred to as the *gram panchayat*[3]). He served as my primary liaison until the end and facilitated introductions to other key individuals in the community. In Sonamukhi, my initial contact was with a tribal gentleman who, whilst not holding an official role in local governance at the time, commanded respect within the village community due to his employment with the government and his past involvement in various social services. Although my interactions with non-tribal families in Sonamukhi were limited in the first three months, arrangements were later made through a local official for me to reside with a non-tribal family. Subsequently, my relationship with the non-tribal families deepened as I became more integrated into the non-tribal community. This close immersion with the non-tribal community provided an opportunity to gain a comprehensive understanding of the power dynamics between tribal and non-tribal groups, which was critical for grasping institutional politics and associated power dynamics in the area.

At the outset of field research, securing suitable accommodation within the villages became my primary concern. Despite the villagers' best intentions, finding a house with an additional room to spare for an outsider proved challenging. Consequently, I accepted the offer extended by the Forest Department to stay at the Forest Rest Hut[4] within the Forest Range complex until a place within the village could be arranged. The Forest Range, a unit of forest administration, constituted the second tier from the bottom in West Bengal and held primary responsibility for implementing forestry schemes. Situated approximately 7–10 km away from the Range Office, getting to my field villages required transportation. Given the absence of public transport, I engaged a car to commute to the villages, typically leaving very early in the morning as life in the heart of the forests was as bustling as that of a city, with activities commencing well before sunrise.

Alongside participant observation, I initiated the process of conducting household surveys whilst becoming acquainted with village life. To ensure timely access to participants, I made a concerted effort to arrive early in the morning. Most villagers set out from their homes early, around 7 am, either tending to their lands or seeking wage labour. Typically, they returned home late in the evening after completing their errands. Men often engaged in chitchats in nearby markets or partook in some recreational activities like playing cards. The bicycle was the most common mode of transport used by men and children alike. Consequently, I frequently extended my stay in the village until late evening to engage with families and conduct household surveys.

On occasions when my arrival in the village was delayed, or on days when I returned during daylight hours, I seized the opportunity to observe the operations of the Range and Beat Offices. The Beat, situated beneath the

Range in the hierarchy, was in close proximity to the Range Office complex. Both offices often remained vacant and tranquil throughout the day, as most staff members, including the Range Officer and Forest Guards, dedicated the majority of their time to the field. There was no designated staff member assigned to remain in the office. Whilst it is logical for most Forest Department staff to spend their time in the field due to the nature of their job, given the collaborative nature of forest governance under joint forest management, it may not be unreasonable to expect visible interaction between the Forest Department and local people in the Range and Beat offices.

The villages, too, assumed quiet and deserted ambiences during the daytime, with only individuals who had some work to do at home or who had young children to look after. During the day, I wandered around exploring the villages and the forests. It was one of these lonely and silent afternoons in Patrasayer that I saw the village council representative riding his bicycle. Upon enquiry, he disclosed his destination – the village council office, and extended an invitation to accompany him. Despite initial hesitations, I accepted the invitation, as I had no pressing engagement. I was initially apprehensive, thinking that my research focus centred on examining the politics surrounding forest protection committees, an autonomous institution with no direct connection with the operation of the village council.

Upon stepping into the village council office, I was immediately struck by its lively and dynamic ambience. The space was abuzz with activity, equipped with state-of-the-art technologies like desktop computers, humming printers and a fast broadband connection. The room reverberated with the orchestrated chaos of governance in motion. Over the subsequent two to three hours, I bore witness to a spectacle of democracy in action. Elected representatives, both women and men, seized the reins of decision-making. They navigated the intricacies of governance, whilst state-appointed officials provided the technical and administrative scaffolding necessary for their endeavours. This experience was a stark juxtaposition to the serene, almost cloistered ambience of the local forest offices, whether at the Beat or Range level. In the ensuing week, subsequent to deliberations, consultations and obtaining necessary permissions from the local village council and other authorities, I resolved to undertake a comparative analysis between the forest protection committee and the functioning of the village council.

Concurrently, I also arranged a space to stay at the local school in Patrasayer and with a non-tribal family in Sonamukhi. However, I primarily stayed with the latter family throughout the fieldwork period due to security concerns raised by local officials. These concerns stemmed from my status as a female researcher and the perceived risk of staying in an isolated school building. I felt compelled to accept this arrangement of uneven time allocation between the two villages, as failure to do so would have jeopardised my ability to continue my research.

In the end, spanning across two villages and encompassing two distinct forms of decentralised institutions, my field research evolved into a multi-site ethnography. The comparative examination of these institutions significantly enriched my research experience. Although gaining entry into the field sites posed challenges, it ultimately proved beneficial to my research.

Covert politics and household surveys

As mentioned, I began the household surveys whilst staying at the Range Office complex and continued the process upon relocating to the villages. Due to the small size of the habitats, I opted to survey all families. However, unlike tightly scheduled interviews, I approached the household surveys within the framework of ethnography, allowing for ample time for interaction (Herbert, 2000). Whilst employing a structured questionnaire, the conversations remained informal and relaxed throughout. The questionnaire served more as a checklist, with the flow of discussion guided by the interviewees themselves rather than being imposed upon them (Silverman, 1985). This approach revealed covert incidents, offering valuable insights into local politics and power dynamics within the village and yielding many leads for my institutional research on distributional and procedural injustices.

During these surveys, I uncovered the existence and significant influence of the traditional tribal council within the village. This revelation stemmed from a family's account where a woman had been labelled as a witch, resulting in the social exclusion of most relatives, including neighbours in her hamlet. Similarly, I learned about instances of families being unjustly excluded from membership in forest protection committees for various fictitious reasons. These issues raised by families deprived of certain benefits emerged as crucial indicators of distributive politics, highlighting the need to explore potential procedural injustices within the operations of forest protection committees and other governing bodies. Upon reflection, I am reminded of Ellen's observation that a fieldworker often 'acquires much valuable information and understanding backstage, when participants remain unaware of the importance of the information being passed on to the researcher' (1984: 145).

It is pertinent to mention that household surveys not only complemented ethnographic inquiry but were also enriched by the ethnographic method. For instance, during the repeat round surveys, ten families in Patrasayer provided detailed information on winter crops, whereas only three families had mentioned these crops in the initial survey. Additionally, a notable disparity in the Total Livestock Unit (TLU) count in Sonamukhi was found between the two livestock holding surveys conducted at the beginning and end of my stay. This significant increase in TLU count, particularly of small stocks like goats, during the later survey was attributed to the underreporting during the initial survey period. The openness of villagers and the tendency to conceal information diminished as I became more integrated into the community, allowing for a deeper understanding of the contrast between verbal accounts and reality through lived experiences and progressive socialisation into the group (Lofland, 1995).

The process of settling down in a research area is often facilitated by local elites, who subsequently take on the role of gatekeepers (Reeves, 2010). Moreover, the close relationships formed with these elites can unintentionally lead to a detachment from the wider community, especially amongst marginalised families. Thus, effectively connecting with these populations and building trust poses a complex challenge (Reeves, 2010). The implementation of household surveys emerged as a crucial strategy in dispelling negative perceptions and cultivating support and collaboration. Furthermore, these surveys facilitated the identification of key individuals within the village for ethnographic interviews.

Ethnography of local institutions

Understanding the effectiveness of institutions in achieving inclusive forest governance posed several challenges, including identifying formal and informal institutions operating within the villages. Additionally, there was the task of experiencing the live processes of institutional functioning, such as decision-making processes, particularly the extent of devolution of power for effective local governance. Moreover, understanding the politics of local institutions entails discerning the contextual factors contributing to inequity and injustice in their operations. Three primary methods proved critical in addressing these objectives: participant observation, document analysis and ethnographic interviews.

Using participant observation for mapping formal and informal institutions

Participant observation is a qualitative data collection method in which the researcher immerses themselves in the environment under study to collect and record data in the form of actions, narrations or numbers (Atkinson and Hammersley, 2007; Musante, 2015; Spradley, 1980). However, Atkinson and Hammersley (2007: 248) contend that observation can be either participant or non-participant, contingent upon whether the researcher assumes an 'established participant role in the scene of study'. Whilst research ethics necessitate researchers to disclose the overarching goals and objectives of the researcher, the efficacy of the method relies on how discreetly the researcher explains the purpose of their research so that participants become less conscious of their behaviour. For example, in examining sensitive topics like gender equality in decision-making processes, by stating that research aims to understand broader processes like the functioning of the village council rather than explicitly focusing on gender dynamics, researchers can create a more natural or unbiased environment for data collection. In essence, whilst it is crucial to hold ethical standards, researchers often become strategic to ensure the validity and reliability of the research findings.

In certain instances, participant observation complements the stated objective of the research. In my original research plan, the objective was to examine the functioning of forest protection committees within the village. This may explain why villagers initially omitted to mention the presence of any other institutions, particularly informal ones. Additionally, in their close-knit society, the significance of long-standing collective activities may be overlooked when interacting with outsiders. Informal institutions often embed themselves within daily practices, operating through everyday practices aimed at prioritising local values such as cooperation and loyalty. Thus, the presence of an informal social regulator like the tribal council went unnoticed during early interactions, until insights gleaned from household surveys revealed the existence of the traditional village council.

By immersing myself in the daily life of the village, I also became aware of the existence of a few informal institutions. Subsequently, further probing with key informants revealed additional informal and formal institutions. For instance, on one occasion, I encountered a bustling crowd near the village common room early in the morning, where I discovered the porch filled with freshly caught fish. Villagers informed me that these fish were caught by members of the *fishing club* from a nearby pond. Membership in this club held significant importance, as it afforded members access to a complementary supply of fish, a staple food and a vital source of nutrition in the area. Whilst there was no formal membership fee, numerous families, particularly those from poor backgrounds, found themselves excluded from this group, consequently being deprived of essential nutrition.

The subsequent exploration with key informants during ethnographic interviews unveiled the existence of additional formal institutions, such as school committees and informal institutions, such as the *elephant society* and *football club*. During my initial visit to Patrasayer, villagers emphasised the expertise of local men in driving elephants away from human habitats. The intrusion of elephants and resulting damage to crops and property in villages close to forests posed a recurring issue in certain regions of the state. However, there was no mention of the collective organisation of these men in the form of a committee until it was disclosed towards the end of my stay. Despite being an informal institution, membership in this committee held significance, as the Forest Department worked through this committee whenever their assistance was needed. Consequently, non-members were excluded from participation and thus forfeited potential remuneration despite possessing the requisite skills.

Through daily observation of village life, I discerned the presence of formal and informal institutions beyond the forest protection committees. Once identified, my interest shifted to understanding the dynamics within these institutions: identifying primary decision-makers, beneficiaries, and the process of membership determination. Central to this investigation was the exploration of whether similar patterns of inclusion and exclusion occurred across institutions, shedding light on the power dynamics. By mapping these patterns, I could identify societal elites and discern contextual factors, particularly socio-political dynamics, influencing access to benefits.

Interviews served as a tool to delve deeper into these contextual factors, providing insights into the mechanisms favouring certain individuals whilst marginalising others from accessing benefits. It became evident that informal groups reflected underlying power dynamics through membership and decision-making processes more than formal institutions. Access to informal collective groups was predominantly controlled by village elites. Thus, informal groups play a crucial role within the community, and studying informal institutions is critical even if the primary focus of research is on any formal institution, as it is the plurality of institutional arrangements that influences socio-political dynamics in any area.

Understanding institutional functioning

Alongside the task of mapping, comprehending the politics of institutional operation is not straightforward. Anthropologists have drawn rich insights into the functioning of institutions in India, such as the office of the land record keeper (Gupta, 1995) and the *gram panchayat* and *anganwadi* (Corbridge et al., 2005), through participant observation. However, studying the activities of most village-level institutions, both formal and informal, including forest protection committees, poses unique difficulties due to the absence of localised designated physical space for their operations. These institutions often operate in intangible spaces (Datta, 2021).

Moreover, observing the activities of forest protection committees proves challenging as their operations often evade direct observation in the absence of any scheduled events such as regular meeting schedules. Despite being formal entities governed by joint forest management guidelines, the lack of routine and structured events impeded my ability to witness decision-making processes first-hand. The primary imperative became remaining vigilant at all times to avoid missing any committee meetings, which occurred sporadically at Range or Beat offices, altogether three over my entire 13 months of fieldwork.

In addition, the homes of committee office bearers frequently served as venues for day-to-day decision-making in the absence of designated office or meeting spaces. Engaging with these office bearers during early mornings and late evenings afforded opportunities to observe certain transactions directly. The small size of the villages facilitated effective tracking of these interactions.

For institutions functioning within physical spaces, such as the *gram panchayats* and Range and Beat forest offices, participant observation provided insight into the dynamics of state–people interaction within decentralised governance structures. Nevertheless, concerns regarding the authenticity of participant observation data persisted. Vidich (1955) argues that the presence of a researcher creates a significant influence, potentially biasing the observed interactions. Thus, in accordance with Atkinson and Hammersley's (1997: 17) assertion that 'the researcher must find some role in the field being studied', I spent extended hours within these offices reviewing various documents (such as minutes of meetings or annual plans) and extracting secondary information from reports, cashbooks and computers. Although my covert intention was to examine the daily interactions amongst elected representatives, government personnel and citizens, overtly, I presented myself as conducting independent research. This strategic approach enabled the observation of natural interactions that might have otherwise remained obscured.

Questions may naturally arise concerning the ethics of employing such a covert strategy in research. I grappled with these ethical dilemmas, knowing that I was navigating a fine line between full disclosure and achieving the research objectives. Yet, without the strategic use of discretion, participants, both state officials and local representatives, would have been conscious, making it difficult to gauge sensitive behaviour like downward accountability. The effectiveness of research outcomes hinged on the delicate balance between ethical considerations and methodological pragmatism.

During meetings, particularly those focused on annual planning and budget allocation, I had the opportunity to delve into the thin line between decentralisation and 'de-concentration' (Ribot, 2004), witnessing how the state-appointed officials wielded disproportionate influence over the decision-making process in the guise of rules and regulations. The elected representatives, with their mandate from the people, sometimes compromised their desires, whilst, in other instances, they remained resilient and manoeuvred ways to advance their objectives.

Observing decision-making processes in both village council meetings and forest protection committees' gatherings allowed me to grasp the comparative nature of power devolution. This comparison also illuminated the impact, in terms of both advantages and limitations, on the integration process of local representatives into the governance system, particularly in joint forest management. Thus, the study of real-time processes through participant observation showed the 'complex interdependencies' (McCall and Simmons, 1969: 2) in organisational and social systems in the study area.

Similarly, conducting formal and informal visits to key officials/power centres such as the Divisional Forest Officer or the Block Development Officer, whether for interviews or to request permission for accessing data or documents, often presented the best and sometimes unforeseen opportunities to observe power dynamics at play during negotiation with common citizens or political representatives. In essence, participant observation straddles a continuum between covert and overt methodologies, typically situated somewhere between these two extreme poles.

Document analysis

Given the challenge posed by the inherent limitations of understanding the dynamics of institutional functioning, especially who are the decision-makers amongst the community leaders/representatives of local institutions, the rules and the process of decision-making for intangible institutions, document analysis becomes an important methodological approach to uncovering institutional procedures. In my research, document analysis focused particularly on reviewing meeting minutes and other pertinent documents related to the operations of these institutions, including the rule book, membership register, election proceedings and bank documents of the forest protection committees. The information gleaned from document analysis also served to mitigate some of the informational gaps encountered during participant observation in the case of institutions like the *gram panchayat* or the self-help groups. Moreover, it prompted further inquiries for ethnographic interviews, a topic I elaborate on in the subsequent section.

Ethnographic interviews

Whilst participant observation offers the opportunity to observe subjects by detaching from the researcher, the interview stands out as a methodological tool for qualitative research, given that the essence of fieldwork lies in interaction (Ellen, 1984). Through interviews, direct contact is established between the researcher and the informants. According to Heyl (2001: 369), an interview is called ethnographic if:

> researchers have established respectful, ongoing relationships with their interviewees, including enough rapport for there to be a genuine exchange of views and enough time and openness in the interviews for the interviewees to explore purposefully with the researcher the meaning they place on events in their words.

In my research, village-level interviews predominantly followed an ethnographic approach typically involving two stages – rapport building and eliciting information (Spradley, 1979). Conducting household surveys acquainted me with the key informants and provided insights into their psychology and social milieu. Most interviews unfolded over multiple sessions rather than as a one-time discussion. I exercised caution in broaching trivial issues such as intra-village conflict or social sanctions, waiting until the 'collaborative climate' had been reached (Woodhouse, 2007: 173).

Intra-village conflicts, as well as social sanctions in rural areas, are often entangled with the rivalry between two political parties or social factions and are frequently exploited as pretexts to marginalise specific families and settle long-standing or old grudges. Hence, addressing such issues need caution. A collaborative climate typically emerges after spending a significant amount of time, when the researcher feels that a friendly relationship or confidence has been developed with interviewees. I often initiated such conversations with less sensitive topics or those unrelated to my research, as a test case to gauge the reaction and establish rapport. Thus, ethnographic interviews afford the opportunity for multiple engagements over the course of research, facilitated by the ongoing contact between interviewees and the researcher.

Interviews were conducted in a semi-structured format to foster participant comfort and allow for greater voice in interaction (Brewer, 2000). Three to four broad themes were listed for each key informant, serving as prompts during the discussions and also functioning as 'aide-memoir' (Richards, 1996). To maintain a natural setting, I refrained from using tape recorders during interviews and instead transcribed the discussions immediately afterwards. Any notable quotations mentioned by interviewees were noted in the notebook.

Ethnographic interviews were conducted with various stakeholders, including village elites, marginalised members, forest committee members and individuals associated with other committees. Local-level officials, such as Forest Guards and other personnel involved in day-to-day forest management, were also interviewed. Village council officials were interviewed to understand their roles and how they resolved differences with elected members, providing clarity on procedural matters.

Interviews in institutional research serve multiple purposes. In the beginning, to gather background insights into the region and the study villages. They also helped to uncover local perspectives regarding procedural justice or injustices, as the notions of justice are embedded within social customs and behaviours. The local narratives crafted through village-level interviews, including insights from local office bearers, community leaders and even poor villagers, wove together the local narratives on decentralised governance. These narratives emerged as drivers in shaping the discourse on decentralised governance in my research. It also prompted a reconsideration of the conventional dichotomy between top-down and bottom-up approaches within the decentralised forest governance framework. In another instance, one of the poorest persons remarked – 'the government has allocated enough provisions to benefit the poor, yet it's the few influential people in the society who reap the benefits, leaving the poorest of the poor stagnant for years to come', prompted reflection on whether grassroots empowerment alone can address the problem of inequity.

In addition to scheduled interviews, several unscheduled interviews were conducted, especially with women who collected fuelwood and other forest produce collectors. These impromptu encounters, whether in forests or at village tube wells, yielded valuable insights into village politics.

It is important to note that various methods, such as participant observation, document analysis and interviews, do not proceed linearly. Instead, they are used simultaneously, with information from one method often prompting queries for another method and vice versa. Thus, ethnographic research is a dynamic process, as illustrated in Figure 8.1.

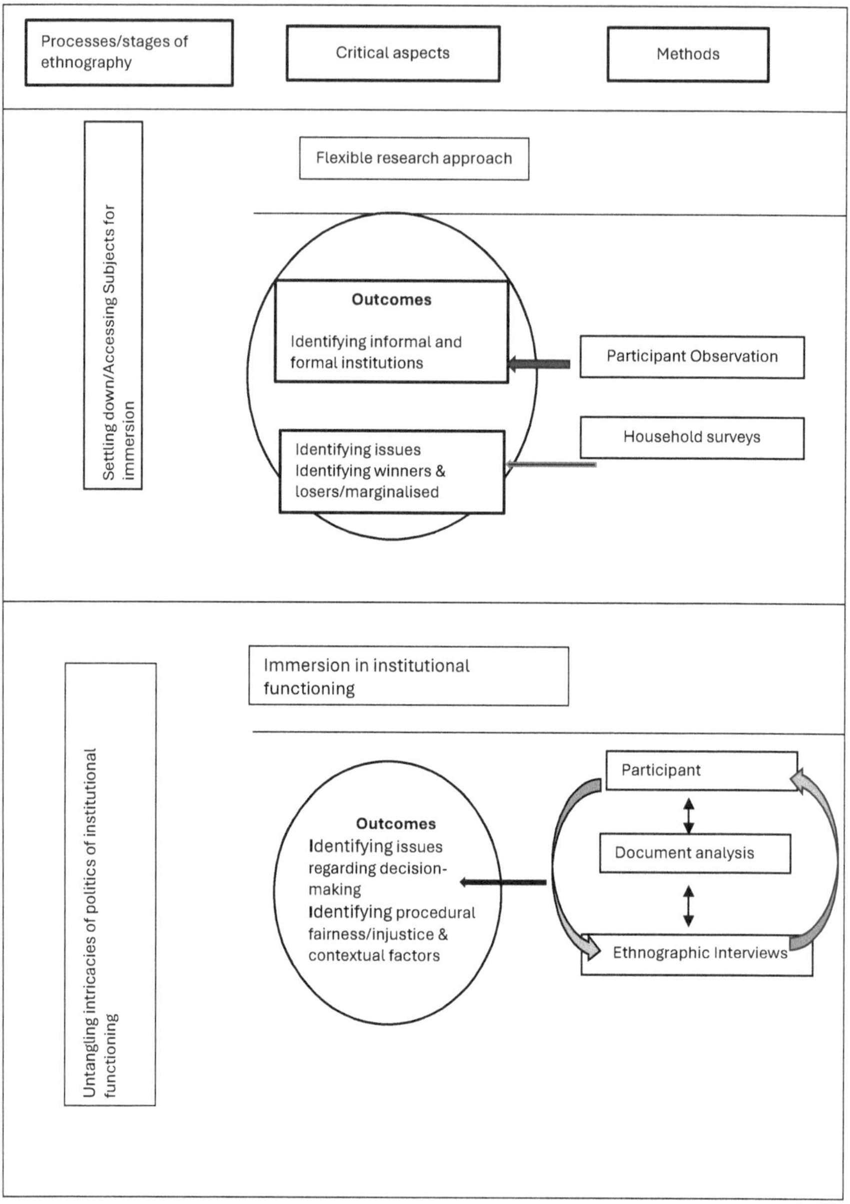

Figure 8.1 An ethnographic approach to studying institutions within joint forest management in India.

Micro–macro linkages – ethnography gives the opportunity to see the process and its influence from changing external events

The commitment to prolonged immersion in a localised setting, inherent in an ethnographic approach, offers the opportunity to observe and assess the macro–micro linkages in everyday life, including the functioning of local institutions. Whilst the local institutions are firmly enmeshed in the local culture and politics, it is crucial to recognise that the socio-political dynamics of different communities and their influence on the local institutions are not static. Instead, the local power game between social groups (such as between the Scheduled Tribes and non-Scheduled Tribes social groups in this case) shifts with political changes at the macro level. Such macro-level shifts can be challenging to observe without an ethnographic approach.

My field research time coincided with the beginning of a shift in political power at the state level. This afforded me the opportunity to witness first-hand how the increasing influence of the opposing political party at the state level affected power dynamics at the village level. In one of the *gram panchayats*, the rival party secured victory in the local election, resulting in a majority of elected representatives being from non-tribal communities. This political transition also led to changes in school staffing within the village, and development projects became focused on non-tribal hamlets. This aspect of village-level power dynamics highlights how macro politics trickle down to a small village and how protagonists in power struggles within small villages take advantage of wider political relationships.

Data recording

In ethnographic research, recording data poses significant challenges, particularly in participant observation and informal conversations. One major challenge I encountered was transcribing interview data without a tape recorder present. As a result, data recording heavily relied on note-taking, which became a continuous and integral aspect of my field methodology. I employed two methods of note-taking: rapid notation of essential observation and information, often in abbreviated or schematic form, and detailed note-taking. Quick notes were typically made during interviews or ongoing observations. Capturing the interviewee's reaction or inquiries regarding my gender and approach to research as a female investigator, thereby fostering opportunities for reflexivity. Additionally, I developed shorthand techniques for efficient note-taking.

Detailed notes were reserved for late evening sessions, allowing for a more comprehensive record of the day's observations and insights. Consequently, the data often represented a convergence of recalled peripheral materials and evoked fringe elements (Schwartz and Schwartz, 1955), offering a selective yet insightful glimpse into the phenomena under study (Hammersley, 1992).

Ethical issues

Ethical considerations played a significant role throughout my field research and data analysis process, as I had to 'reconcile the rights and interests' of research partners of diverse backgrounds, such as informants, gatekeepers, marginalised villagers, local governments and others (Ellen, 1984: 137). Recognising this, I obtained informed consent from all major participants in the research. Firstly, I obtained consent from the state Forest Department in the state capital, as the forests in the study area were entirely government-owned. Secondly, since I decided to include the functioning of the *gram panchayat* in my study, I also obtained consent from the *panchayat* offices and from the block development office.

The most critical issue was explaining my research to the participants of the study villages and gaining their support. All questionnaires, including the household questionnaires and the checklist for semi-structured ethnographic interviews, were translated into Bengali, the local language. Additionally, I provided information about myself and my study on the first page of the questionnaire and mentioned the anonymity and confidentiality of information. In the rural setting, the other major issue was to manage villagers' expectations of me. I explained to them that my research was not connected to granting schemes or benefits but was purely academic. Moreover, during my stay in the villages and whilst participating in village events or meetings, I remained a passive observer.

Ethical considerations come to the forefront during the use of data for publications, as this process involves sharing observations, analysis and conclusions derived from a long-term association with the research site, potentially accessible to individuals within the research communities. This introduces added pressure on the researcher, as they contemplate how research participants might react to even seemingly trivial observations. To mitigate this impact, I anonymised the identities of all research participants in the analysis, thereby safeguarding against any potential harm or unintended consequences to any individual or community that may arise from research findings.

Summing up

The case study reveals that exploring the intricacies of institutions is a multifaceted task, beginning with the daunting task of unravelling their structural intricacies, encompassing actors, systems and regulatory frameworks. This challenge is compounded when the institution operates in intangible space, as exemplified by the forest protection committee within joint forest management in India, and most informal institutions operating in the rural landscapes. Within such contexts, actors assume diverse roles contingent upon their positions. Whilst some individuals occupy overt decision-making roles, others function as recipients of institutional actions. Certain actors wield conspicuous influence within the realm of decision-making, whilst others, notably local elites, clandestinely manipulate proceedings.

Moreover, assessing the efficacy of institutions like the forest protection committee and analogous local bodies engaged in natural resource management necessitates an evaluation of equitable benefit distribution amongst beneficiaries. Consequently, identifying beneficiaries emerges as a formidable task in itself. Ethnographic immersion in village life facilitates nuanced interactions across societal strata, enabling the identification of both beneficiaries and marginalised stakeholders within the institution.

Participant observation emerges as a pivotal methodology for comprehensively grasping institutional dynamics, both formal and informal, elucidating the complexities of decision-making processes, including how and where the decisions are made. Document analysis serves as an invaluable tool for delineating institutional systems, regulations, and the roles assumed by primary actors.

Importantly, field research culminates in a series of follow-up ethnographic interviews, elucidating critical issues pertinent to decision-making processes and benefit allocation. The iterative nature of institutional study necessitates multiple research cycles, with ethnography's protracted engagement enabling researchers to revisit and refine their inquiries over time. Lastly, a flexible approach to research, including a flexible research design, is critical for gaining maximum outcomes from the ethnographic research approach. It is important to keep patience as constraints often turn into opportunities.

Notes

1 The names of both villages were replaced with the names of the Blocks.
2 Renamed as Ministry of Forest, Environment and Climate Change in 2017.
3 Rural India is governed by a three-tier, democratically-elected structure known as the *Panchayati Raj Institute*. The *gram panchayat*, which consists of a group of villages with electorates of 10,000–12,000, is the lowest tier of the *panchayat* structure.
4 It was an ensuite accommodation situated next to the Forest Range Office (FRO). It is generally used by the officers for short rest, and so there was no support staff.

References

Atkinson, P. and Hammerseley, M. (2007) *Ethnography: Principles in Practice*, 3rd edition, New York: Routledge.

Baviskar, A. (1996) *In the Belly of the River*, Delhi: Oxford University Press.

Brewer, J.D. (2000) *Ethnography*, Buckingham: Open University Press.

Brockington, D. (2002) *Fortress Conservation: The Preservation of the Mkomazi Game Reserve*, Tanzania, Oxford: James Currey.

Cleaver, F. (2002) Reinventing institutions: Bricolage and the social embeddedness of natural resource management, *European Journal of Development Research*, 14(2): 11–30.

Cleaver, F. (2012) *Development through Bricolage: Rethinking Institutions for Natural Resource Management*, London: Earthscan.

Corbridge, S., Williams, G., Srivastava, M. and Veron, R.E. (2005) *Seeing the State: Governance and Governmentality in India*, New York: Cambridge University Press.

Datta, S. (2021) Making decentralisation work: A comparative ethnographic analysis of forest conservation and village governance in West Bengal, India, *Conservation and Society*, 19(4): 205–217. https://doi.org/10.4103/cs.cs_20_81

Ellen, R.F. (ed.) (1984) *Ethnographic Research. A Guide to General Conduct*, London: Academic Press.

Gupta, A. (1995) Blurred boundaries of corruption, the culture of politics, and the imagined state, *American Ethnologist*, 22(2): 375–402.

Gupta, A. and Ferguson, J. (1997) *Anthropological Locations: Boundaries and Grounds of a Field Science*, Berkeley: University of California Press.

Hammersley, M. (1992) *What's Wrong with Ethnography? Methodological Explorations*, London: Routledge.

Herbert, S. (2000) For ethnography, *Progress in Human Geography*, 24(4): 550–568. https://doi.org/10.1191/030913200100189102

Heyl, B. (2001) Ethnographic interviewing, in Atkinson, P., Coffey, A., Delamour, S., Lofland, J. and Lofland L. (eds.) *Handbook of Ethnography*, London: Sage, pp. 369–383.

Kraidy, M.M. (2002) The global, the local, and the hybrid: A native ethnography of glocalization, in Taylor, S. (ed.) *Ethnographic Research: A Reader*, London: Sage Publications, pp. 187–210.

Li, T.M. (2014) *Land's End: Capitalist's Relations on an Indigenous Frontier*, Durham, NC: Duke University Press.

Lofland, J. (1995) Analytic ethnography: Features, failings and future, *Journal of Contemporary Ethnography*, 24: 30–67.

Manor, J. (2004) User committees: Potentially damaging second wave of decentralisation, *European Journal of Development Research*, 16(1): 192–213.

McCall, G.J. and Simmons, J.L. (eds.) (1969) *Issues in Participant Observation: A Text and Reader*, Reading, MA: Addison Wesley Publishing Company.

McDermott, M., Mahanty, S. and Schreckenberg, K. (2013) Examining equity: A multidimensional framework for assessing equity in payment for ecosystem services, *Environmental Science Policy*, 33: 416–427. https://doi.org/10.1016/j.envsci.2012.10.006

Murphy, E. and Dingwall, R. (2001) The ethics of ethnography, in Atkinson, P., Coffey, A., Delamont, S., Lofland, J. and Lofland, L. (eds.) *Handbook of Ethnography*, London: Sage, pp. 339–351.

Musante, K. (2015) Participant observation, in Bernard, H.R. and Gravlee, C.C. (eds.) *Handbook of Methods in Cultural Anthropology*, Lanham, MD: Rowman & Littlefield, pp. 251–292.

Reeves, C.L. (2010) A difficult negotiation: Fieldwork relations with gatekeepers, *Qualitative Research*, 10(3): 315–331. https://doi.org/10.1177/1468794109360150

Ribot, J. (2004) *Waiting for Democracy: The Politics of Choice in Natural Resource Decentralisation*, Washington, D.C.: World Resources Institute.

Richards, D. (1996) Elite interviewing: Approaches and pitfalls, *Politics*, 16(3): 199–204.

Sandström, C. (2009) Institutional dimensions of co-management: Participation, power, and process, *Society & Natural Resources*, 22(3): 230–244. https://doi.org/10.1080/08941920802183354

Schwartz, M.S. and Schwartz, C.G. (1955) Problems in participant observation, *American Journal of Sociology*, 60(4): 343–354. http://dx.doi.org/10.1086/221566
Silverman, D. (1985) *Qualitative Methods and Sociology*, Aldershot: Gower Publishing.
Spradley, J.P. (1979) *The Ethnographic Interview*, New York: Holt, Rinehart and Winston.
Spradley, J.P. (1980) *Participant Observation*, New York: Holt, Rinehart and Winston.
Sunam, R.K. and McCarthy, J.F. (2010) Advancing equity in community forestry: Recognition of the poor matters, *International Forestry Review*, 12(4): 370–382.
Vidich, A.J. (1955) Participant observation and the collection and interpretation of data, *Amererican Journal of Sociology*, 60(4): 354–360. https://www.jstor.org/stable/2772028
Woodhouse, P. (2007) People as informants, in Thomas, A. and Mohan, G. (eds.) *Research Skills for Policy and Development: How to Find Out Fast*, London: Sage Publications, pp. 159–179.

9 Staying with the trouble

Experimenting with a socio-technical approach to researching water governance

Tavengwa Chitata, Frances Cleaver and Jeltsje Kemerink-Seyoum

Introduction

Local water governance is formed in the interaction between institutional and infrastructural processes. For example, infrastructural configurations (canals, gates, pipes, taps) can facilitate or impede the functioning of institutional rules and arrangements (relating to accessing, sharing, protecting water) and vice versa. Researching these processes together requires us to pay significant attention to operationalising an approach in which neither element is privileged nor excluded from data collection and analysis. Critical institutionalism has advanced our understanding of the functioning of socially embedded institutions, particularly in understanding the practices, processes and meanings (Cleaver, 2002). In this regard, critical institutionalism has shown the importance of routinised practices and behaviours in natural resources governance, including water (Chitata, Cleaver and Kemerink-Seyoum, 2023; Cleaver, 2002, 2012; de Koning, 2014; Faggin and Behagel, 2018; Kemerink et al., 2018; Van der Kooij, Zwarteveen and Kuper, 2015). In addition, the functioning of institutions governing water has been explained through processes of bricolage. In such processes, people select and blend different elements (rules, roles, norms) from existing social arrangements, to create arrangements that fit their particular water management needs (Cleaver, 2002). Critical institutionalism and bricolage analysis have furthered understanding of how blended institutions emerge and endure, with a focus on the ways in which claims to legitimacy and authority help to gain recognition for them (Cleaver and de Koning, 2015). This involves an appreciation of the historical context and the dynamic unfolding of institutions over time. People constantly interpret, bend and adapt the rules, behaviours and practices, sometimes with changes to the meanings attributed to them, according to circumstances (Cleaver, 2002).

Critical institutionalism has primarily relied on ethnographic qualitative methodologies focusing on the social phenomena and rules as the starting point for inquiry. In-depth interviews, participant observation and immersion in the case studies, group discussions, and open-ended and semi-structured questionnaires are commonly used tools for data collection. For example, in researching the drivers of institutional change in the commons in Burkina

DOI: 10.4324/9781032624679-9

Faso, Karambiri et al. (2020) used in-depth interviews and focus group discussions. Marin and Bjorklund (2015) relied on document review, interviews and participant observation to explore institutional dynamics and land tenure in Finland, while Faggin and Bahagel (2018) used interviews and site visits to research forest management in Brazil. De Koning (2014) used interviews, questionnaires, participant observation and group exercises to understand institutions in practice in the governance of forests in Bolivia and Ecuador. On the other hand, mainstream institutionalism has relied more on quantitative methods including questionnaire surveys, statistical analysis and in some cases use of controlled lab experiments to test structural variables (Ostrom, 1993, 2010). The starting point for most mainstream institutional analysis has been *(adherence to)* formal rules and rights. Of note here are studies including those done by Ostrom on rules and rights in irrigation schemes using large data sets and meta-analysis of irrigation schemes (Ostrom, 2010, 1993). In addition, Mukuyu et al. (2024) in their study on groundwater in the Kavango-Zambezi Transfrontier Conservation Area traced the frameworks and laws for polycentric groundwater governance. Lam and Chui (2016) in their study of institutional nesting and robustness applied a combination of in-depth interviews and appraisal surveys, while Baggio et al. (2016) undertook a meta-analysis of 69 cases related to irrigation, fisheries and forestry.

These methodological approaches translate into the separation of research on institutions into the social dimensions of the phenomenon on the one hand and the performance and contestations of rules and technology (or infrastructure) on the other (Hoogesteger et al., 2023; see also Pinch, 2008). These methodologies and analyses of institutions have shaped longstanding debates about the relative importance of agency and structure in explaining people's behaviour; and about how people make choices in the governance of natural resources. This has had implications for our understanding of how progressive social change happens or does not happen and how power relations and inequalities are maintained, contested and transformed.

Most research in critical institutionalism and mainstream institutionalism shares a methodological commonality – privileging the scrutiny of formal and informal rules and arrangements. Yet, the rules and arrangements are often enabled or constrained by infrastructures like boreholes, pumps, pipes and canals. For example, despite conducting research on many irrigation schemes, a number of authors have remained preoccupied with analysing the rules and pay less attention to the influence of infrastructure in the institutional dynamics (Baggio et al., 2016; Hoogesteger, 2015; Lam and Chui, 2016; Ostrom, 1993; Skjølsvold, 2010). Some proponents of critical institutionalism have evolved their research methods and analysis to include the materiality of infrastructure in shaping the rules in use and water governance arrangements (Benouniche, Zwarteveen and Kuper, 2014; Chitata, Kemerink-Seyoum and Cleaver, 2021; Kemerink et al., 2019; Kuper et al., 2017; Kuper, Mayaux and Benmihoub, 2023; Mirhanoğlu et al., 2023; Silva-Novoa Sanchez, Kemerink-Seyoum and Zwarteveen, 2019). We propose to advance this growing socio-technical approach to researching institutions by experimenting with technography. In our research, this has involved trying out different ways of understanding the entanglement of material and institutional arrangements; of taking infrastructure seriously without neglecting the rules and norms of pieced-together institutions.

Technography has been defined as 'study of and immersion in technology in practice/use – which is the observation, description, and study of infrastructure and/or technologies and their use, change and [agency] within a specific social and historical context' (Chitata, 2022: 14). This socio-material approach has its traditions in Science and Technology Studies (STS), as well as being associated with feminist scholarship dating back to the 1960s. It was developed as an attempt to show that technology and social relations are co-constituted; how power and inequalities are associated with infrastructures and technology; and how infrastructures are constantly being re-shaped through people's practices. (Anand, 2023; Haraway, 2003, 2023; Jansen and Vellema, 2011; Pinch, 2008; Zwarteveen et al., 2024). Using technography as an approach helps us to acknowledge the complexity of water governance problems, and the entanglement of human, material and non-human entities, without being limited to a narrow mono-disciplinary or methodological approach. We borrow the phrase 'staying with the trouble'[1] to characterise this approach, with the intention of embracing interconnections, contradictions, experimentation and uncertain outcomes in the research focus and process.

Although a technographical approach has been around for a long time, and has been developed conceptually, for example in relation to agricultural sustainability (Arora and Glover, 2017), its operationalisation for data collection and fieldwork is not fully developed (exceptions include Jansen and Vallema, 2011; Kemerink et al., 2019; Silva-Novoa Sanchez, Kemerink-Seyoum and Zwarteveen, 2019). In this chapter, we present how we experimented with the operationalisation of technography to understand the working of institutions and water governance arrangements in irrigated agriculture, using a case of the Rufaro Irrigation Scheme in Zimbabwe. In the following section, we present the context in which technography was operationalised and the positionality of the researcher which influenced the outcomes of the method. This is followed by an explanation of technography in practice, that is, how it was operationalised in the Rufaro Irrigation Scheme, insights generated using this approach and concluding remarks.

The context and positionality

This operationalisation of technography and the insights were part of the first author's PhD research in the Rufaro Irrigation Scheme, Zimbabwe (see Chitata, Kemerink-Seyoum and Cleaver, 2021, for a description of the Rufaro Irrigation Scheme). The method was implemented during the fieldwork of over two and half years, with a cumulative 15 months of ethnographic immersion in the Rufaro community. The data collection began before the

COVID-19 pandemic in 2019 and continued during the pandemic up to October 2021 (see Chitata, 2022). In this chapter, we alternate between 'I' and 'we' in the different sections of the chapter mainly because this work is the primary research of the first author (I), yet it also represents a collaborative effort (we) between the first author and PhD supervisors – who were involved in the development of the research, mentoring and co-publishing papers as part of the PhD studies at the University of Sheffield. The first author did the data collection and was in charge of operationalizing the technography approach.

I, Tavengwa Chitata, was initially trained and practiced as an irrigation engineer. In line with my training, both as an irrigation engineer and later on as a social scientist, I am shaped by two epistemological traditions, and so is my research approach, including the tensions and different approaches this marriage brought. Frances Cleaver is a hybrid social scientist and a proponent of Critical Institutionalism and Jeltsje Kemerink has both engineering and social sciences training, like the first author. As a team, we share a common interest in challenging binary classifications (of concepts, methodologies, analysis) and a commitment to working with the complex inter-relationships and indeterminacy that we see in many human–environment interactions.

The choice of technography is an attempt to 'stay with the trouble' of reconciling my hybrid academic training in engineering and social sciences (Haraway, 2016: 116) but also reflects the supervisory team. As a team we were able to draw on our strengths in different disciplines to consider the ways in which insights from engineering, political ecology, STS and critical water studies were complementary or contradictory. This might have been difficult if I had trained in or was supervised by a team from one single discipline. Operationalising an interdisciplinary approach meant integrating my knowledge and understanding of irrigation system design, construction, repair and maintenance with critical institutionalism and critical water studies, as well as talking to and observing farmers in their everyday irrigation practices. In addition, I grew up in a farming community in Zimbabwe, where I participated in agricultural activities, including irrigation. This experience was valuable for implementing technography, communicating with farmers, and sometimes sharing relatable experiences.

My dual engineering and social science backgrounds have shaped my belief in the situatedness of knowledge production and the rejection of single, objective truth. The research methods and choices made were never neutral or apolitical. As a researcher, I came into this research with my ideology, convictions and experiences, which then influenced the outcome of the research process. The same applies to my supervisors – we had to 'stay with this trouble' and navigate our differences or similarities. We shaped each other based on our understanding of the world and co-shaped some of the choices I made in this research. My understanding of the world was also co-shaped by my interactions with the farmers involved in the research process (see also Given, 2008).

Research and its outcomes are a social process co-shaped and co-produced by the actors involved. For example, ideologically, I believe in the promotion of justice in water governance with a sensitivity to gender issues. This shapes how I observe the practices of women farmers and their interactions with the watery spaces. In this research, paying attention to how women interacted with infrastructure allowed us to see how knowledge and institutional arrangements around groundwater were gendered. These gendered relationships were shaped by men's and women's different experiences of access to different infrastructures (Chitata, Kemerink-Seyoum and Cleaver, 2021). My beliefs and orientation also had implications for how I stepped in to help women who were marginalised. For example, some members of the community saw me as an impartial outsider who they could confide in on the politics of land and water in the irrigation scheme, and who could help find recourse for injustices. Tracing the changes in the irrigation infrastructure over time also entailed tracing the changes in the land sizes and locations of irrigators within the irrigation scheme. In tracing land sizes over time and relating land size to change in the infrastructure, it became apparent that some farmers had bigger plots than others. When bringing these observations into discussion with irrigators it was noted that there was some 'land grabbing' happening within the irrigation scheme. In particular, some men were increasing their land sizes by stealth through incremental extension of their field boundaries into other farmers' plots. In this case, the operationalisation of technography uncovered these irregularities and brought them to be part of the conversations with farmers. This led to a request from the women farmers and irrigation management committee to mediate long-standing land disputes within the irrigation scheme. These were plot boundary conflicts usually involving women and men. On most occasions' men were encroaching into land belonging to women neighbours who were often either of old age or widowed. The irrigation committee had tried solving these issues without much success as the accused would refuse to implement the remedial action and accuse them of favouritism. With the help of a land surveying application – GPS field area measure – on my phone, I managed to show that other farmers had indeed encroached into their neighbours' land and readjusted the boundaries. Interestingly, when data generated by this technology was shown to the farmers, it was not disputed. This also shows how tools of technographical research can be used in mediating conflicts and how being an objective non-participant is difficult and possibly unhelpful as immersion in a research may lead to participation in the research setting.

Staying with the trouble for this research also involved being reflexive in the research process. This meant striking a balance and reflecting all the time, particularly on the influence I had as an active participant in the research process and the need to detach and take a step back from the moments of the phenomena but also acknowledging it is not possible all the time (see also Given, 2008). For example, I was actively involved in the irrigation community's everyday activities during the fieldwork. On one occasion, I was involved in helping out with repairing a broken borehole: providing my labour, passing on tools and offering suggestions when the farmers were brainstorming how to carry out the repairs. I was wholly involved in the process I was studying and researching at the same time – participating, shaping and contributing to a pre-existing arrangement for infrastructure repair. Being involved also means going beyond the physical presence to emotional engagement – sharing the pains and joys of events within the community.

As part of the reflexivity exercise for this research, I constantly reflected on and made known in my thesis how my critical stance was shaped by the economic, social and political repression in Zimbabwe, which was more intense after 1998. Reflexivity was one of the most critical aspects of the research, constantly asking myself why I had a particular view and trying not to privilege any knowledge – engineering over social science and vice versa. From my perspective, and as part of the exercise of reflexivity, I tried to understand a variety of forms of knowledge and logics of action. All these are equally crucial in informing how institutions function, progress and change (see also Charmaz, 2017; Denzin and Lincoln, 2011).

I was also aware of, and acknowledged, the existence of inequalities within the community, such as gender, seniority in the irrigation scheme, and age. Yet, I tried not to reify or encourage the perpetuation of these inequalities. This also meant amplifying in my (and our) writing the voices of the marginalised and the ways they were exercising their agency. It is important to note that the question of human/non-human agency was part of the trouble encountered by the research team and we had to stay with this trouble, and it was something that we had to continuously negotiate yet it remained not entirely resolved.

My gender also had implications in my research. Being male, attached to the Department of Irrigation (previously and for this research) and doing my studies abroad meant that I was given some form of status within the community. I had to be conscious of this position and reflect on how this was affecting my research and shape how the farmers responded to me. Being male allowed me to have access to public spaces like the local beer hall at any time and interact with men in the community with ease. However, as a male, I could not visit women farmers in their homes because, culturally, I was a stranger and was not supposed to have 'private' interactions with women in the community. All the in-depth interviews with women were conducted in the irrigation scheme, while interviews with men were conducted in the comfort of their homes or the irrigation scheme, depending on their preference. As a former member of the Department of Irrigation, which is responsible for the rehabilitation of the irrigation scheme, I experienced a certain level of mistrust in the initial stages of the research. The farmers felt I was evaluating the wrongs they were doing to a newly installed infrastructure, and I was monitoring their response to government-sanctioned programmes like command agriculture.[2] It took many rounds of explanation to convince the farmers that I was there just for research, and, through daily interactions, I built trust and good relations with the community.

Operationalising technography: infrastructure as an entry point

Taking non-human objects – in this case infrastructure – seriously entailed for me decentring focus from the human beings and taking infrastructure as the object of analysis and the entry point of the research. How did I approach the infrastructure or where to start? After entering the community and doing all the formalities needed to be given access to the community spaces and infrastructure, I started off with a document review of the history of the infrastructure in the Rufaro Irrigation Scheme. I did this through a search on infrastructural project reports, feasibility documents, aerial photographs, Google Earth, rural district planning maps and documents, and irrigation system design and rehabilitation documents from the Department of Irrigation. These documents covered 35 years.

A participatory mapping exercise followed this for which I organized transect walks together with fifteen (15) of the 55 members of the irrigation scheme. Of the group of irrigators who joined for the participatory mapping and transect walk, eight (8) were volunteers, and the community chose seven (7) based on their specific knowledge of the community water infrastructure and/or being in the irrigation management committee. The participatory mapping exercise with the community members aimed to map all the water infrastructures in the Rufaro community and consolidate and triangulate with the information gathered through the document reviews. Furthermore, I used the mapping exercise to identify places and moments in which infrastructure was changed. To further understand the institutional processes, practice and meanings facilitated by the infrastructure in the Rufaro Irrigation Scheme, we mobilised socio-technical tinkering as a methodological approach to delve into the history and evolution of the infrastructures (see also Archidiacono et al., 2024; Kemerink et al., 2019; Leonardelli, Kemerink-Seyoum and Zwarteveen, 2023; Quimby et al., 2023; Silva-Novoa Sanchez, Kemerink-Seyoum and Zwarteveen, 2019). Socio-technical tinkering in the context of our research on the Rufaro Irrigation Scheme refers to 'acts that produce deviations from the initial plans and designs of water infrastructure in terms of the form, materiality and functioning of the infrastructure' (Chitata, 2022: 21). These acts can vary in space and time and nature and magnitude – they range from ad-hoc/informal and minor adjustments to the infrastructure to make water flow to the different parts of the irrigation scheme to planned/formal large-scale rehabilitation events (Chitata, Kemerink-Seyoum and Cleaver, 2021; Leonardelli, Kemerink-Seyoum and Zwarteveen, 2023). In this case, we observed both minor changes, such as a small incision in the old canals, and much larger changes, such as replacing a canal system with a pressurised system. Therefore, socio-technical tinkering involves continuous 'fine-tuning of infrastructure, adaptations, negotiation, adjustments' and individual and collaborative efforts (Archidiacono et al., 2024: 1). Smaller and larger changes in the infrastructure configurations over time and rehabilitations were captured and documented, together with the functioning of the infrastructure and events that triggered

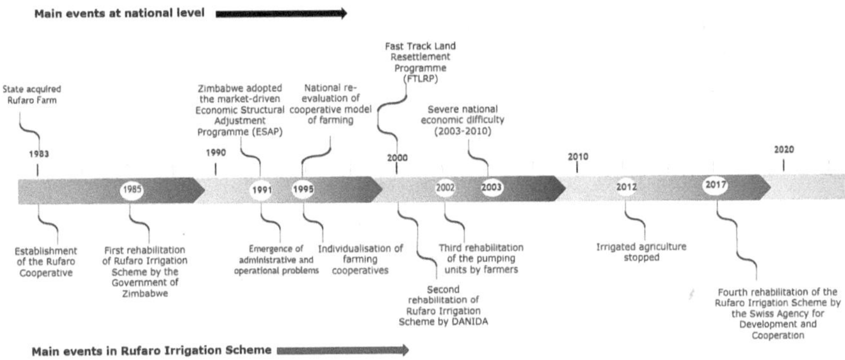

Figure 9.1 Depiction of major historical socio-technical tinkering practices on infrastructure (combining infrastructural and institutional elements).

or influenced some of these changes. Figure 9.1 gives a graphic overview of the history of the infrastructure of the Rufaro irrigation scheme.

Linking the material and the social

The observed changes in the infrastructure shaped the conversations with farmers and engineers – conversations in which we explored the reasons, processes and implications of these modifications on water governance arrangements. The technography orientation facilitated the investigation of how people interacted with the infrastructure, how they understood it, how they changed it and how it changed them. The ethnographical approach allowed us to enrich the observed changes, modifications and practices of irrigators on infrastructure with thick descriptions of the everyday social interactions, history, cultural reasoning behind their actions and everyday functioning of the infrastructure. This was further complemented by observing irrigators' everyday practices inside and outside the irrigation scheme. As highlighted in my positionality, I was a participant observer 'seeing the words in practice' (Chitata, 2022: 16). Being a participant observer enabled me to connect with the farmers and familiarise myself with them more humanely – experiencing the emotions and physical demands of irrigated agriculture. For example, I would join the irrigators when they were irrigating their fields at night and help out – being in that moment with them. The farmers had to do irrigation at night because the electricity for pumping water to the irrigation scheme was back on in the evening as part of the load shedding by the Zimbabwe Electricity Supply Authority (ZESA). As a participant observer, I attended monthly and ad-hoc meetings and participated in agricultural activities, operation and maintenance of the infrastructure, cattle dipping and watering and social events in the community. I would make a conscious decision to assume a particular role.[3] The roles ranged from being a complete observer, a participant as an observer, an observer as a participant and a complete participant (see also Given, 2008).

Respondents and data collection tools

Researching institutions or doing fieldwork can never be fully planned and I had to stay with the trouble of planning the fieldwork yet not knowing what will happen in implementing technography. The data collection methods and tools evolved in the field and were adjusted over time depending on the stage of the research yet also affected by the changing circumstances, particularly the COVID-19 pandemic (see Figure 9.1). The combined techno-ethnographical approach allowed me to collect very visible manifestations together with thick descriptions of how people engage with water and irrigation infrastructure through several rounds of interviews with both farmers (44, of which 23 men and 21 women) and government officials (12 in total, including engineers involved in design and maintenance of the infrastructure). The farmers were selected based on a stratified random sampling technique to ensure representation of different gender and age groups and to include those who farm in the different parts of the irrigation scheme (see Figure 9.1). The interviewed government-employed engineers were all men, in the absence of any female engineers working on this irrigation system and were selected using convenient sampling based on easy accessibility and availability. This data was complemented with, and triangulated by, focus group discussions (4), diaries (6)[4] and participant observations, participatory mapping activities (8)[5] and a thorough desk study of relevant documents and literature.

Participant observation was not always planned and no specific actors were chosen for observation but it would take place in the spaces where irrigation activities would be happening. Participant observation was key in this research, as it helped me to understand and triangulate some of the interviews – seeing the words in practice – and also to have a deeper understanding of the physical, emotional and psychological meanings attached to activities, as they were experienced by the individuals involved. For example, it was only through participating in cleaning the water night storage tank that I could feel the drudgery involved, which I could only imagine from a participant's description of it as 'hard labour' and as 'care' work. As a participant observer, I was involved in monthly and ad hoc meetings on irrigation as well as being involved in irrigating and harvesting crops, weeding, operation and maintenance of the infrastructure, cattle dipping and watering and social events in the community. Using the technographical method, I produced detailed descriptions of different irrigation infrastructures, their historical roots, the modifications made through time, how they were operationalized, the everyday practices of the actors involved in using and changing them as well as the broader socio-technical processes involved. Data collection was continued up to a pragmatic[6] point of data saturation on some elements. This was a combination of reaching a point where no new information was coming out from further interviews or technographic engagement with the irrigation infrastructure and the pragmatic difficulty of endlessly continuing to interview people about ever-changing circumstances with the limited time allocated for PhD research.

Data analysis: dialogical model

The data analysis for the case study was a dialogical process involving understanding the case study in its own right and an open engagement between the data and the concepts to think through the empirical evidence (see also Lund, 2014; Rule and John, 2015). The data was also dialogical within itself in an incremental manner, in the sense that the data collection did not follow the linear process of conducting fieldwork and then going back to the office to analyze all the data as one set. Instead, the data collection and data sets emerged over the two and half years and were punctuated by episodes of analysis and paper writing – together with my supervisors – as highlighted in the timeline for data collection (Figure 9.2). Thus, these data sets were dialogically feeding into each other, providing cumulative and incremental meaning. I analysed the collected data using thematic analysis (Given, 2008). This was achieved through thematic coding created by repeatedly reading interview transcripts (Babbie and Mouton, 2001; Green et al., 2007). The data sets were compared to identify similarities and differences and identify relationships and networks within and between data collected at different stages. The data from written interviews were thoroughly read to identify themes and subthemes. I used three types of codes: descriptive, thematic and analytical. The descriptive code explains the attributes of the data source – that is, the interviewee's attributes and relations with other interviewees. The themes are the issues that emerged while doing the research, either topics that interviewees brought up and captured in the interview narratives or from (grey) literature, while the analytical code was used to make sense of the data by linking it to concepts mobilized for this research. In coming up with the themes, I engaged with the data – thorough reading – and, in the process, making a nuanced understanding of the interview narratives, asking questions such as what are the main ideas and concerns being communicated by the interviewee (open coding) and developing categories that capture these main issues. Because of the nature of the process of data collection, analysis and write-up (Figure 9.2), the open coding would be done on the next set of data and cleaned to avoid duplication by making comparisons with the other parts as well as adding new themes and making new connections between themes, thus connecting different narratives through similar themes and categories – that is, axial coding (Rule and John, 2015). The interview data were triangulated to ensure rigour in the data analysis through crosschecking with other data sources, including participant observations, GIS mapping and flow measurement in the irrigation scheme (see also Maher et al., 2018).

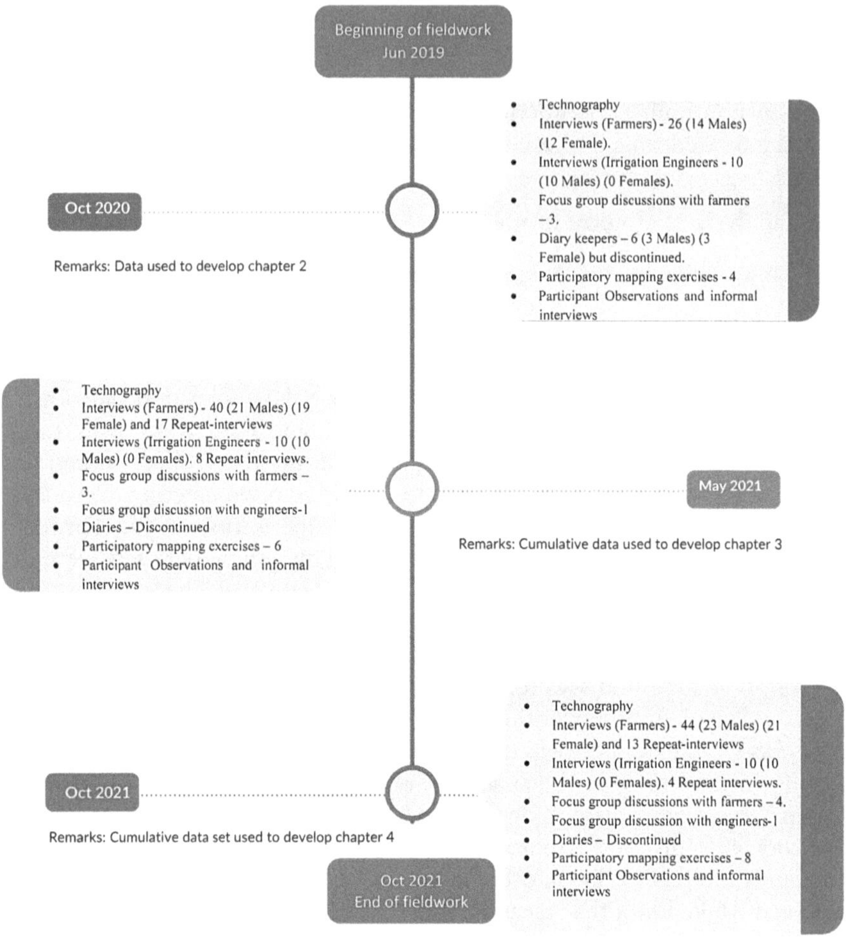

Figure 9.2 Evolution of data collection process, analysis and write-up.

Findings: everyday socio-technical tinkering and institutional arrangements

The socio-material interactions are always entangled with institutions. The planned and everyday small adjustments to infrastructure shape and are shaped by the water governance arrangements. In this section, we present a snippet of findings from the operationalisation of technography in the Rufaro Irrigation Scheme. The last major infrastructural change (to date) in the Rufaro Irrigation Scheme took place between 2016 and 2018 with a change from surface irrigation system to pressurised system (see Chitata, Kemerink-Seyoum and Cleaver, 2021 for the details about the change). In the case of the Rufaro Irrigation Scheme, a change in the materiality of water infrastructure – from the lined canals requiring collective action to manage distribution to pressurised surface irrigation where water is supplied through pipes and hydrants serving individual fields – changed how the downstream and upstream irrigators accessed water. Before the change, upstream irrigators had better access to irrigation water from the canals and hence a privileged spatial position which translated to a social power – the downstream irrigators needed to negotiate with the upstream to get water for irrigation. After the infrastructural change, the downstream farmers had better access to irrigation water, and with this change, they also acquired a privileged position to access water first (Chitata, Kemerink-Seyoum and Cleaver, 2021). In the era of lined canals, successful irrigation largely depended on collaboration. However, introducing pipes and hydrants aligned to individual plots changed the dynamics of cooperation and dependency in the irrigation scheme. This change meant that each farmer had a hydrant on their plot.

Irrigation engineers were anticipating that having each farmer with a hydrant supplying water to their plot will give more access and control to the farmers to irrigate their crops whenever needing water without much negotiation with other farmers or the Irrigation Management Committee. The use of these hydrants presented an opportunity for the farmers to access water at any time yet it presented a big problem to the Irrigation Management Committee and, in particular, the people who were responsible for allocating water. It became difficult to restrict water access for the people who had not paid their monthly contribution to the irrigation scheme – contribution towards the purchasing of electricity to pump water. In response to this, the Irrigation Management Committee decided to remove all the levers on the ball valves fitted to the hydrants with the intention of giving the levers to only those irrigators who had paid their monthly fees and were on their turn for irrigating. This shows that the change in the infrastructure required more institutional efforts to manage and arrange for water access thus showing the hydro-socio-material entanglements in the irrigation scheme.

This arrangement was not well received by the irrigators – those who were up to date with their payments and in arrears alike. In response, the irrigators would bring a pair of pliers to force open the valves on the hydrant and irrigate. However, the use of pliers resulted in the wearing-off of threads on the valves to an extent that the lever which was supposed to be used for opening the valves on the hydrants by the Irrigation Management Committee would no longer fit or be useful. These everyday practices of using infrastructure together with the institutional efforts to control everyday access to water for irrigators by the irrigation management committee resulted in the unintended consequences to the infrastructure. Although the use of pliers and adjustments on the hydrants by the individual irrigators may seem minor pragmatic practices, multiplied over many hydrants the effects became systemic – the irrigation scheme was only semi-functional and it would cost substantial amounts of money to repair. Collecting such funds would fall to the Irrigation Management Committee and undoubtedly be conflict ridden and costly in terms of time and social goodwill.

In this example, we have highlighted three critical points explored using and paying attention to socio-technical tinkering of infrastructure: the importance of infrastructure in mediating social relations of power; the changes in infrastructure require more institutional efforts to modify and rethink water governance arrangements; and the unintended consequences of the everyday infrastructural and institutional adjustments.

How does technography approach inform and enrich critical institutionalism?

Critical institutionalism is sufficiently able to analyse rules and processes, power dynamics and the working of institutions from the social phenomena perspective; technography can enrich this approach by showing the intricate and evolving relationships between infrastructure and the social phenomena. In particular, a technography approach shows that infrastructures are about more than their functionality, but also shape social relations of power and water governance practices. Furthermore, it helps to illuminate the unintended consequences of socio-technical tinkering with infrastructure and how this can result in institutional change and (re)adjustments and (re)negotiation of water governance arrangements. Lastly, bringing technography into engagement with critical institutional is not a seamless, mutually enriching process. In the team, we all recognize the entanglement of the socio-material world and, therefore, agreed about the added value of putting infrastructure more central in our analysis, yet we differed in our views as to how to theorize these socio-material relations and whether, for instance, it could be useful for analytical purposes to understand infrastructure as having agency. 'Staying with the trouble' also meant for us that this difference remained unresolved.

Conclusion

In this chapter, we have shown how we carried out an experimental and interdisciplinary socio-technical approach to studying institutions governing water and irrigation infrastructure in the Rufaro Irrigation Scheme in Zimbabwe. Technography illuminates how the form and materiality of water infrastructure are vital in shaping interactions and facilitating transformations between humans and non-human entities. The approach accounts for changes in social relations of power within irrigated agriculture as an outcome of changes in the form and nature of infrastructure, thus explaining structure and agency beyond the social realms.

By presenting this approach, it is by no means the only method that can be used to explore socio-material entanglements yet we find it useful in enriching critical institutionalism as it enables disciplinary methods to be extended and do interdisciplinary research to understand institutional change and functioning.

Operationalisation of technography is our attempt at valuing the different human and non-human aspects of water governance, and the contributions that social science and engineering can make to enriching critical institutionalism. This does not mean technography is neutral and apolitical. The process of reflexivity is still key in keeping live the politics and 'staying with the trouble' of researching institutions. Throughout, the process is influenced and shaped by our backgrounds, ideologies and the community in which the research is carried out. We end this chapter by encouraging more research to bring a technographical approach into engagement with institutional analysis in order to enrich critical institutional scholarship.

Notes

1 'Staying with the trouble' is a phrase from Donna Haraway's book titled 'Staying with the trouble Making Kin in the Chthulucene'. In this chapter it means acknowledging the complexity of solving problems and recognising the limitations of any disciplinary perspective and complexity of knowledge production. Furthermore, it also means being experimental and unsure of outcomes as well having contradictions within oneself.

2 Command agriculture was a government program imposed on smallholder farmers to grow maize and winter wheat with support from the government on the understanding that the smallholder farmers will sell their produce to the state grain reserve – Grain Marketing Board (GMB).

3 This was particularly practical during my first months in the community, with time I to some degree lost this autonomous choice. The more I became integrated the more the community sometimes would determine my role. For example, the elderly saw me as their child whom they could task to do some things or I would be asked to give an opinion or help with deciding during a meeting which I had planned to be 'complete' observer.

4 The keeping of diaries was discontinued in the first round of data collection as they caused tensions amongst the irrigators. A section of the irrigators did not view the resources given for diary keeping as only for data collection but as a material gain for the other as I supplied them with pens and books for diary keeping. Furthermore, the majority of the farmers were not writing their diaries and for those who did, the big chunk of the information I would have observed or heard through participant observation. Also, when COVID-19 started in Zimbabwe, I left the field briefly and the farmers substituted writing their diaries with telling me during update calls.

5 Participatory mapping involved transect walks around the village with a group of irrigators to explore a phenomenon or locating historical points, irrigators participating in recreating the irrigation infrastructure as drawings or represent their village and relative locations of everything using representative material like stones and twigs among others.

6 This is pragmatic in the sense that complete data saturation is in practice not achievable as there are always some new details that will be coming out in several rounds of interviewing people and also the infrastructural configuration and adjustments are a continuing process. This is shaped by the changing conditions and circumstances during the research period. Rather than it being totally planned as at reaching the saturation point it becomes to some extent artificial and guided by the time there is to do fieldwork.

References

Anand, N. (2023) Anthroposea: Planning future ecologies in Mumbai's wetscapes, *Environment and Planning D: Society and Space*, 41(4): 683–706. https://doi.org/10.1177/02637758231183439

Archidiacono, S., Kemerink-Seyoum, J.S., Leonardelli, I., Dominguez Guzman, C., Chitata, T. and Zwarteveen, M. (2024) Engineering as tinkering care: A rainwater harvesting infrastructure in Cochabamba, Bolivia, *Engineering Studies*, online first. https://doi.org/10.1080/19378629.2024.2304176

Arora, S. and Glover, D. (2017) Power in practice: Insights from technography and actor-network theory for agricultural sustainability, *STEPS Working Paper 100*, Falmer: STEPS Centre, University of Sussex.

Babbie, E. and Mouton, J. (2001) *The Practice of Social Research* (South African edition), Cape Town: Oxford University Press.

Baggio, J.A., Barnett, A.J., Perez-Ibara, I., Brady, U., Ratajczyk, E., Rollins, N., ... and Janssen, M.A. (2016) Explaining success and failure in the commons: The configural nature of Ostrom's institutional design principles, *International Journal of the Commons*, 10(2), 417–439. https://doi.org/10.18352/ijc.634

Benouniche, M., Zwarteveen, M. and Kuper, M. (2014) Bricolage as innovation: Opening the black box of drip irrigation systems, *Irrigation and Drainage*, 63(5): 651–658. https://doi.org/10.1002/ird.1854

Charmaz, K. (2017) The power of constructivist grounded theory for critical inquiry, *Qualitative inquiry*, 23(1): 34–45. https://doi.org/10.1177/1609406917719350

Chitata, T. (2022) *Living with Flows of Groundwater and Infrastructure Governance: The Case of Rufaro Irrigation Scheme, Zimbabwe*, PhD thesis, Sheffield: University of Sheffield.

Chitata, T., Cleaver, F. and Kemerink-Seyoum, J. (2023) 'Our humanism cannot be captured in the bylaws': How moral ecological rationalities and care shape a smallholder irrigation scheme in Zimbabwe, *Environment and Planning E, Nature and Space*, 6(4): 2761–2780. https://doi.org/10.1177/25148486221137968

Chitata, T., Kemerink-Seyoum, J.S. and Cleaver, F. (2021) Engaging and learning with water infrastructure: Rufaro irrigation scheme, Zimbabwe, *Water Alternatives*, 14: 690–716.

Cleaver, F. (2002) Reinventing institutions: Bricolage and the social embeddedness of natural resource management, *European Journal of Development Research*, 14(2): 11–30. https://doi.org/10.1080/714000425

Cleaver, F. (2012) *Development through Bricolage: Rethinking Institutions for Natural Resource Management*, London: Earthscan.

Cleaver, F. and de Koning, J. (2015) Furthering critical institutionalism, *International Journal of the Commons*, 9(1): 1–18. https://doi.org/10.18352/ijc.605

de Koning, J. (2014) Unpredictable outcomes in forestry—Governance institutions in practice, *Society & Natural Resources*, 27(4): 358–371. https://doi.org/10.1080/08941920.2013.861557

Denzin, N.K. and Lincoln, Y.S. (2011) *The Sage Handbook of Qualitative Research*, 4th edition, Thousand Oaks, CA: Sage.

Faggin, J.M. and Behagel, J.H. (2018) Institutional bricolage of Sustainable Forest Management implemented in rural settlements in Caatinga biome, Brazil, *International Journal of the Commons*, 12(2): 275–299. https://doi.org/10.18352/ijc.872

Given, L.M. (ed.) (2008) *The Sage Encyclopedia of Qualitative Research Methods*, Thousand Oaks, CA: Sage.

Green, J., Willis, K., Hughes, E., Small, R., Welch, N., Gibbs, L. and Daly, J. (2007) Generating best evidence from qualitative research: The role of data analysis, *Australia and New Zealand Journal of Public Health*, 31(6): 545–550. https://doi.org/10.1111/j.1753-6405.2007.00141.x

Haraway, D. (2003) A manifesto for cyborgs. Science, technology and socialist feminism in the 1980s, in Blaikie, A., Hepworth, M. and Holmes, M. (eds.) *The Body: Critical Concepts in Sociology* (Vol. 1), London: Routledge.

Haraway, D. (2016) Staying with the trouble: Anthropocene, capitalocene, chthulucene, in Moore, J.W. (ed.) *Anthropocene or Capitalocene: Nature, History, and the Crisis of Capitalism*, Oakland, NC: PM Press.

Haraway, D. (2023) When species meet, in Franklin, A. (ed.) *The Routledge International Handbook of More-than-Human Studies*, London: Routledge.

Hoogesteger, J. (2015) Normative structures, collaboration and conflict in irrigation; A case study of the Píllaro North Canal Irrigation System, Ecuadorian Highlands, *International Journal of the Commons*, 9(1): 398–415. https://doi.org/10.18352/ijc.521

Hoogesteger, J., Konijnenberg, V., Brackel, L., Kemink, S., Kusters, M., Meester, B., Sanjeev Mehta, A., 't Hart, T., van der Poel, M., van Ommen, P., Boelens, R. and Sanchis-Ibor, C. (2023) Imaginaries and the commons: Insights from irrigation modernization in Valencia, Spain, *International Journal of the Commons*, 17(1): 109–124. https://doi.org/10.5334/ijc.1216

Jansen, K. and Vellema, S. (2011) What is technography?, *NJAS-Wageningen Journal of Life Sciences*, 57(3–4): 169–177. https://doi.org/10.1016/j.njas.2010.11.003

Karambiri, M., Brockhaus, M., Sehring, J. and Degrande, A. (2020) 'We are not bad people'-Bricolage and the rise of community forest institutions in Burkina Faso, *International Journal of the Commons*, 14(1): 525–538. https://doi.org/10.5334/ijc.1061

Kemerink-Seyoum, J.S., Chitata, T., Guzmán, C.D., Novoa-Sanchez, L.M. and Zwarteveen, M.Z. (2019) Attention to sociotechnical tinkering with irrigation infrastructure as a way to rethink water governance, *Water*, 11(8): 1670. https://doi.org/10.3390/w11081670

Kemerink-Seyoum, J.S., Tadesse, T.M., Mersha, W.K., Duker, A.E.C. and De Fraiture, C. (2018) Sharing benefits or fueling conflicts? The elusive quest for organizational blue-prints in climate financed forestry projects in Ethiopia, *Global Environmental Change*, 53: 265–272. https://doi.org/10.1016/j.gloenvcha.2018.10.007

Kuper, M., Benouniche, M., Naouri, M. and Zwarteveen, M. (2017) 'Bricolage'as an everyday practice of contestation of smallholders engaging with drip irrigation, in Venot, J.P., Kuper, M. and Zwarteveen, M. (eds.) *Drip Irrigation for Agriculture*, London: Routledge, pp. 256–265.

Kuper, M., Mayaux, P-L. and Benmihoub, A. (2023) The persistent appeal of the california agricultural dream in North Africa, *Water Alternatives*, 16(1): 39–64.

Lam, W.F. and Chiu, C.Y. (2016) Institutional nesting and robustness of self-governance: The adaptation of irrigation systems in Taiwan, *International Journal of the Commons*, 10(2): 953–981. https://doi.org/10.18352/ijc.638

Leonardelli, I., Kemerink-Seyoum, J. and Zwarteveen, M. (2023) Obliqueness as a feminist mode of analysing waterscapes: Learning to think with overflows,

Environment and Planning E: Nature and Space, 6(2): 863–879. https://doi.org/10.1177/25148486221117725
Lund, C. (2014) Of what is this a case?: Analytical movements in qualitative social science research, *Human Organisation*, 73(3): 224–234. https://doi.org/10.17730/humo.73.3.e35q482014x033l4
Maher, C., Hadfield, M., Hutchings, M. and De Eyto, A. (2018) Ensuring rigor in qualitative data analysis: A design research approach to coding combining NVivo with traditional material methods, *Intenational Journal of Qualitative Methods*, 17(1): 1–13. https://doi.org/10.1177/1609406918786362
Marin, A. and Bjørklund, I. (2015) A tragedy of errors? Institutional dynamics and land tenure in Finnmark, Norway, *International Journal of the Commons*, 9(1): 19–40. https://doi.org/10.18352/ijc.586
Mirhanoğlu, A., Özerol, G., Hoogesteger, J., Van den Broeck, P. and Loopmans, M. (2023) Socio-material Bricolage, *International Journal of the Commons*, 17(1): 69–86. https://doi.org/10.5334/ijc.1188
Mukuyu, P., Nyambe, N., Magombeyi, M.S. and Ebrahim, G.Y. (2024) Polycentric groundwater governance: Insights from the Kavango-Zambezi transfrontier conservation area, *International Journal of the Commons*, 18(1): 322–336. https://doi.org/10.5334/ijc.1336.
Ostrom, E. (1993) Design principles in long-enduring irrigation institutions, *Water Resources Research*, 29(7): 1907–1912. https://doi.org/10.1029/92WR02991
Ostrom, E. (2010) Beyond markets and states: Polycentric governance of complex economic systems, *American Economic Review*, 100(3): 641–672. https://doi.org/10.1257/aer.100.3.641
Pinch, T. (2008) Technology and institutions: Living in a material world, *Theory and Society*, 37: 461–483. https://doi.org/10.1007/s11186-008-9069-x
Quimby, B., Nichols, C.M., du Bray, M.V., Cantor, A., Bausch, J.C., Wutich, A., and Brasier, K. (2023) Changing flows: Sociotechnical tinkering for adaptive water management, *Environmental Management*, 71(2): 421–431. https://doi.org/10.1007/s00267-022-01744-2
Rule, P. and John, V.M. (2015) A necessary dialogue: Theory in case study research, *International Journal Qualitative Methods*, 14(4): 1–11. https://doi.org/10.1177/1609406915611575
Silva-Novoa Sanchez, L.M., Kemerink-Seyoum, J.S. and Zwarteveen, M. (2019) Water infrastructure always in-the-making: Distributing water and authority through the water supply network in Moamba, Mozambique, *Water*, 11(9): 1926. https://doi.org/10.3390/w11091926
Skjølsvold, T.M. (2010) One size fits all? Designer-institutions: Lessons from two flawed attempts in Malawi, *International Journal of the Commons*, 4(2): 758–771. https://doi.org/10.18352/ijc.118
Van der Kooij, S., Zwarteveen, M. and Kuper, M. (2015) The material of the social: The mutual shaping of institutions by irrigation technology and society in Seguia Khrichfa, Morocco, *International Journal of the Commons*, 9(1): 129–150. https://doi.org/10.18352/ijc.539
Zwarteveen, M., Domínguez-Guzmán, C., Kuper, M., Saidani, A., Kemerink-Seyoum, J., Cleaver, F., ... and Bhat, S. (2024) Caring for groundwater: How care can expand and transform groundwater governance, *International Journal of the Commons*, 18(1): 384–396. https://doi.org/10.5334/ijc.1350

10 From rules to practices

Assembling natural resource governance in the Peruvian Amazon

José Carlos Orihuela

Introduction

The opening chapter of this collective volume argues that theory and methods of institutional analysis are assorted and that much of such variety is due to the significance of contextual conditions. In particular, it states that 'historical context and existing institutions matter and may affect how a new institution is received, performs, and is shaped over time and space' (Nunan, Chapter 1). This chapter has in this observation a cornerstone: institutional change is complex and contingent. In essence, mainstream institutionalism has a narrow understanding of the significance of historical and ecological contexts, oversimplifies power dynamics and politics, focuses on equilibria rather than change and all too commonly assumes that formal rules matter more than, or can be studied independently from, organizational dynamics, culture and ideas. This has been discussed extensively (Abers and Keck, 2013; Campbell, 2004, 2005; Cleaver, 2012; Evans, 2004; Orihuela, 2018; Orihuela et al., 2021). Acknowledging the limitations of mainstream institutionalist literature, however, does not imply neglecting its validity altogether. Instead, I advocate for the critical use of conceptual tools and theoretical frameworks.

This chapter introduces an assemblage perspective to explore institutional diversity in natural resource governance. An assemblage is defined as the evolving convergence of human and non-human actors, their actions and their environments. Resource governance either comes together through governmental practices or fails to do so. Then, a central argument posits that practices are consistently more important than rules. This assertion is empirical rather than normative, a lesson drawn from resource governance studies in developing countries (Abers and Keck, 2013; Orihuela et al., 2021). How does the strength of the state or alternative non-state organizations in enforcing formal 'rules of the game' fluctuate when juxtaposed with the influences of cultural tradition and economic power? What novel forms of resource governance emerge from the interplay of new and established actors of diverse economic, political, cultural and epistemic authority? Though it always depends on where and when these questions are posed, the ultimate institutional determinant of economic behaviour lies not in written rules but in actual practices.

DOI: 10.4324/9781032624679-10

Formal rules matter if and when they become practice, that is, when there is a sufficiently strong organization to enforce them. However, even in the presence of a state or an alternative Leviathan, well-established or in the making, some form of bricolage occurs (Cleaver, 2002, 2012; Cleaver and Whaley, 2018). Given that any organization governing natural resources operates in a social context, a given rule's translation into various resource governance practices varies across space and over time, influenced by the local social and institutional fabric (Campbell, 2004, 2005). This observation is particularly relevant for countries with extensive and diverse societies, such as Peru and its Amazonian Indigenous peoples, who have historically lacked access to state and formal rule-making yet effectively control significant subnational territories (Blackman et al., 2017; Chavarría, Rummenhöller and Moore, 2020; RAISG, 2020; Smith and Pinedo, 2002). Imported ideas and practices of resource governance travel and get mixed with local ones; conversely, local practice might get exported, transformed into and hybridized with globally diffusing formal rules.

Acknowledging bricolage is just the starting point, though. Political economy and political ecology research show that power structures condition the passage and implementation of natural resource rules and policies. To govern nature, the state apparatus advocating for sustainable resource use or ecosystem conservation must contend with the state apparatus that opposes such goals. This struggle within the state unfolds in a multi-layered political arena (Abers and Keck, 2013; Hochstetler and Keck, 2007; Orihuela, 2017, 2020). Moreover, at forest peripheries, environmentally-minded states with varying territorial reach also interact with economic actors, strong or weak, seeking to commodify or self-use land and resources. This spans from forest products to underground minerals, operating under informal rules of greed and need (Orihuela and Mendieta, 2021). Economic actors contesting and influencing, or capturing, formal resource governance can range from transnational corporations and globalized mafia networks to marginalized Indigenous peoples; such diverse actors are interconnected in market, political and social webs (Li, 2007, 2014; Li and Semedi, 2021; Tsing, 2005).

In the realm of natural resource governance, whether pursued by the well-intentioned or the unscrupulous, rules and practices need to be orchestrated through a political economy to hold significance. In the Amazon of Peru, in particular, resource governance is customized by legal and illegal, ever-expanding small-scale mining, timber extraction or coca and palm oil plantation economies. The Peruvian state struggles to effectively control its extensive subtropical forests (Álvarez-Berríos, L'Roe and Naughton-Treves, 2021; Furumo and Aide, 2017; Vijay et al., 2018).

In essence, the transformation of Andean Amazon forests into palm oil/cacao/coca plantations or gold mining sites often unfolds under the informal 'law of the jungle', dictated by the strongest economic power in the field. However, informal and illegal governance may eventually become formal or hybridize with the existing state. This may occur, for example, when 'land use change' from forestry to monoculture agriculture is retroactively legalized or when mining permits are extended to already operating illegal gold miners (Dammert, 2015, 2017; Damonte, 2018). In 2024, without proper debate, and bypassing law-making regulations, the Peruvian Parliament approved significant modifications to the Forest Law (Law 29763) that, according to civil society observers, eases deforestation and land conversion in the Amazon (Infobae, 2024; SPDA, 2023b).

In areas where resource governance remains unsettled, conventional institutional analysis is of limited help. Particularly in the resource peripheries of peripheral countries,[1] where the state diverges from the models theorized by Max Weber or assumed by Douglass C. North, neither rules nor practices of resource governance are stable but, rather, fluid, contested and subject to change. In multi-scalar arenas, a multitude of actors with various political economic motivations seek to govern the peripheral forests of the world economy. The making of community forestry, extractive reserves for sustainable forest use, or REDD-type of ecosystem service conservation projects represents formal governmental initiatives competing with alternative, contentious, informal and illegal ones. Institutional analysis of Amazon's resource governance cannot ignore the contentious and unbalanced nature of this political economy.

Why the Peruvian Amazon?

My choice to research resource governance in the Amazon was motivated by real-world, pressing questions as a Peruvian who had become a global citizen after ten years abroad. By the early 2010s, gold mining was flourishing in the Peruvian Amazon. While preparing class material for undergrad students in the United States, I learned that about one-fifth of the Peruvian Amazon had protected area status, the first national park in the rainforest dating to 1973. Were these new institutions for Amazon conservation capable of halting deforestation? Were they at least active, supported by some critical network of local and transnational actors, shaping new green economies?

Moreover, I also learned that, in the southern Peruvian Amazon, the buffer zone of the Tambopata extractive reserve neighbours the newly extended gold fields, and the protected area was portrayed as exemplary of Indigenous people's participation. I wanted to understand what 'participatory conservation/development' meant in practice, and how Indigenous peoples perceived this contemporary global policy paradigm.

My interest in the political economy of development was further peaked by the fact that illegal gold mining was partly an unintended consequence of a developmental intervention. Years before, the Peruvian state had paved a controversial road connecting Madre de Dios (Peru) with Acre and Rondonia (Brazil), one of those mega projects that the crony and developmental government of Brazil had been financing in the region until the *Lava Jato* mega anti-corruption investigation intervened. The road had been opposed by a loose coalition of conservationists but was welcomed by locals, as public demonstrations showed (Jacopo, 2022; Massi and Nem Singh, 2018).

One research project at a time, I envisioned potentially writing an economic and environmental history of the southern Peruvian Amazon, a biodiversity hotspot and lung of the planet, shaped by its 20th-century commodity cycles, from rubber to gold, and the developmental state intervention.

For non-Amazonian Peruvians like myself, the growing interest in the Amazon and its Indigenous peoples was also a consequence of recent tragic state violence. With Peru's macroeconomy benefitting from the commodity super cycle, political elites had advocated opening the Amazon to large-scale investment and commodification, even using military force if necessary (Acuña 2015; Drinot, 2011). In June 2009, conflict over legislation aiming to ease land trade and private investment in the Amazon escalated, leading the state to send forces to exercise the theorized legitimate use of force. People mobilized against the new laws, blocking roads and the facilities of Petroperú, state-owned oil enterprise operating wells and a pipeline in the northern Amazon. Despite the Peruvian state having signed the ILO Convention 169 on free, prior and informed consent of Indigenous peoples, there had not been proper consultation on the legislation with Indigenous peoples, which angered contentious grassroot leaders. But the government wanted to send a clear message about who had the power to set the 'rules of the game'. The Massacre of Bagua claimed the lives of 39 Peruvians (Hughes, 2010; Santos-Granero and Barclay, 2011).

Why the assemblage framework?

I came to use an assemblage framework in my research because my encounter with resource governance in the southern Peruvian Amazon was marked by its fluid and multifaceted nature. However, I arrived at such an approach only after my initial fieldwork. When I went to carry out research in Madre de Dios, in 2013, I was not looking for Foucault or Latour in the Amazon forests. Coming from an eclectic graduate education focused on economics and the political economy of development, I was certainly thankful but frankly fatigued with the new institutional economics, and so looking for knowledge elsewhere. I knew the questions I had could not be answered by searching for some underpinning universal mechanism and writing overly-rationalistic academic discourse. I believed the political economy had to grapple seriously with history, contingency and agency. But I had not arrived at an assemblage theory framing yet.

As mentioned earlier, the site of my initial project was the Tambopata National Reserve in Madre de Dios, the southern Peruvian Amazon. My loose research questions aimed to uncover how this protected area originated, how it had evolved, particularly in relation to Indigenous peoples, and whether it effectively contributed to the emergence of a new local *green* economy. I wanted to know whether this was just a 'paper park' or if it was an organization with institutional properties, that is, to some degree capable of structuring everyday economic life. To answer these questions, I conducted archival research and open-ended interviews with key actors involved in the invention and operation of the protected area. The goal was to craft a case study exploring how new participatory conservation institutions emerge and evolve. The underlying framework that guided me was my personal assemblage of historical and sociological institutional analysis (Campbell, 2004; Mahoney and Thelen, 2010; Scott, 2001), which I have since regularly worked with (Orihuela, 2014, 2020).

It was after fieldwork that I became fascinated with governmentality and Actor-Network Theory (ANT) readings. I wasn't altogether rejecting my structuralist views on the economy and society to embrace a radically different point of departure for present and future research. I found both literatures offered valuable concepts and new perspectives to pragmatically incorporate into my research. Neither provides a handbook for conducting social science research; in fact, they maintain a certain scepticism, if not cynicism, toward the enterprise (Law, 2004; Latour, 2005; Winkel, 2012). On the one hand, I borrowed the notion that government is the attempt to shape human conduct by calculated means and that governmentality, 'the conduct of conduct', occurs through discourse (Dean 2009; Fletcher, 2010; Foucault, 2008; Li, 2007). On the other hand, I embraced the idea that the social world can also be the constant re-make of webs and associations. Therefore, conventional social science is ill-suited to study 'the ephemeral, the indefinite and the irregular' (Law, 2004: 4). Mark Bevir's concept of 'situated agency' governmentality (1999, 2010) helped round ideas for my first Tambopata article (Orihuela, 2017).

Two important things happened between my first and second research projects on Amazonian protected areas. First, I revisited anthropology of development books that I found insightful, in particular Tania M. Li (2007)'s *The Will to* Improve, Anna L. Tsing (2005)'s Friction and James Ferguson (1994) s *The Anti-Politics Machine.* Second, I collaborated with Arturo Mendieta, an economist and anthropologist conducting ethnographic work in the area. In Orihuela (2017), the main actors were the planners and commanders of the protected area. I had not been able to conduct in-depth interviews with forest-level environmentality crusaders, such as park guards and local environmentalist civil society leaders, or observe much about governmental conservation practices in situ. Arturo and I found in *The Will to Improve* to be a source of inspiration for our research, and that the next Tambopata paper would follow park guards, their networks and their actions both inside and outside the conservation area. My colleague spent five months engaged in ethnographic work in Tambopata National Reserve and the neighbouring Bahuaja Sonene National Park, seeking to understand the governmental practices in place. He became a biological monitoring intern and met and engaged with diverse local actors, including those not following new resource governance, such as Indigenous peoples discontent with nut harvest extractive rules, and local miners, also Indigenous peoples.

The keyword in Orihuela and Mendieta (2021) is assemblage. Drawing on 'practices of assemblage' (Li, 2007) and ANT studies (Latour, 2005, 1988; Law, 2004; Law and Mol, 2008), we conceptualize an assemblage as the dynamic convergence of human and non-human actors, their actions and their environments, in complex webs of relations. This way, we think of governmental rules and practices as part of malleable and complex webs of relations, pulled by specific actors, blending and conflicting with alternative assemblages and sources of power. Environmentalities, that is, governmentalities that have environmentalist goals, exist in a multitude and interact with one another (e.g., Asiyanbi, Ogar and Akintoye, 2019; Choi, 2020; Cortes-Vazquez and Ruiz-Ballesteros, 2018; Palacios and Sarmiento, 2021). Drawing from Li (2007), the government operates by shaping desires and configuring habits, aspirations and beliefs. However, such a government needs to be constructed. Resource governance actors put together legitimacy, technical expertise and a wide range of resources, intervening in and interacting with political-economic structures and ecological systems.

In essence, for resource governance to matter, it must be put into practice, performed and enacted. Resource governance in the Amazon is diverse and dynamic. Contested. Indefinite and irregular. Multiple and hybrid. Everything but stable and uniform. An analytical framework has to help in portraying and explaining such complexity. Or so we decided while designing the research project and writing 'One and three forests: Understanding institutional diversity in Amazonian protected areas'.

Context and power in the evolution of rules

Let's revisit written rules and mainstream institutionalism. As seen in Chapter 1 of this volume, the very definition of institutions remains unsettled. Mainstream approaches conceptualize institutions as 'the rules of the game in a society' that 'structure incentives in human exchange, whether political, social, or economic' (North, 1990: 3). According to the canonical work of Douglass C. North, formal rules are those 'written'; informal rules, instead, encompass conventions and codes of behaviour. A significant shortcoming of mainstream institutionalist research for the study of natural resources is its tendency to overlook context, ignoring that institutional formation is ad-hoc due to bricolage mechanisms (Cleaver, 2002, 2012).

Equally problematic is to ignore power asymmetries and politics. The 'community', in particular, can become the ultimate black box in the institutional analysis of natural resource governance. In mainstream literature, 'community' commonly emerges as a fairly homogenous and harmonic social construct, its members presumed to share the common interest in conserving natural resources (Cleaver, 2012; Hall et al., 2014). Indigenous peoples, in turn, tend to be portrayed as self-governed and deeply connected to land and forests (Li, 2014).

The Indigenous communities we have researched in the southern Amazon of Peru are in fact legally recognized by the state since the 1970s as 'communities of native peoples'. Contrary to the generalizations listed above, however, the 'natives' belonging to a single community are far from homogenous. In the Tambopata forest, different ethnic groups and kin coexist, engaging in different forest economies, both within and outside the protected area, from gold mining to nuts harvesting and ecolodge tourism. Resource governance interests within Tambopata Indigenous communities are diverse and far from harmonic (Orihuela, 2017; Orihuela and Mendieta, 2021).

In turn, these more or less heterogeneous Indigenous communities compete with other social groups over forest use. Commodity cycles of the 20th century involved rubber, gold, oil and timber, each sponsored by the state through tax breaks, subsidies, land property rights, and basic infrastructure. Thus, as state-driven conservationism gained traction in the 1970s, it had to fight state developmentalism forging cattle and plantation economies, extensively promoted in the neighbouring Amazon of Brazil (Dourojeanni, 2009, 2018). To be sure, conventional state developmentalism, sectoral economic policy unconcerned with deforestation and environmental consequences, did not finish back then (Alvarez and Naughton-Treves, 2003).

At the risk of stating the obvious, resource/environmental governance is political and the political does not begin nor end at the community level. Every single protected area established in the southern Peruvian Amazon is the outcome of contentious, multi-scalar, and layered political processes, in which Indigenous peoples have at times gotten together and contested or compromised with governmental conservation projects (Moore and Chicchón, 1997; Orihuela, 2020). For instance, the 'communal reserve', co-managed by the state and Indigenous people organizations, is a protected area model that emerged from prolonged struggles, involving diverse social groups. It originated in response to the expansion of the oil industry and involved a compromise among conservationists and indigenists against the shared threat of extractives (Álvarez et al, 2008; Palacios and Sarmiento, 2021; Smith and Pinedo, 2003).

It is imperative to place context and power at the core of the institutional analysis of institutional resource governance in the Amazon. The mere approval of new 'written rules' tells us little about how natural resources are effectively governed. This isn't to say that the evolution of formal rules is of little academic interest or practical importance. Arguably, good governance in the Peruvian Amazon benefits from written rules that may initially remain dormant, only to become a resource for activism in state and society down the road (Manrique and Orihuela, 2024). In short, the making of new rules needs to be read along with the social context and the political process in which they are embedded.

Which rules structure everyday economic life?

North (1990: 3) defines institutions as guides for human interaction, providing a structure to everyday life. In the case of formal institutions, drawing on North's definition, the rules that matter are the ones that are enforced or, more precisely, those translated into practice by some organization (Campbell, 2005). The practiced matters over the formally written. This nuance of referring to *the practiced* instead of *the enforced* is central. Much of the political science literature seems to assume that the passing of new rules inevitably leads to their enforcement (Abers and Keck, 2013). Similarly, recent literature focused on Latin America examines enforcement gaps and the associated 'institutional strength', defined as formal rules that are stable and enforced (Brinks, Levitsky and Murillo, 2019; Levitsky and Murillo, 2009).

However, governance organizations do not function as straightforward enforcers of formal rules. Belief systems and cognitive mechanisms mediate the translation of rules into practices, as shown by the new institutionalism in organizational sociology (Bromley and Powell, 2012; Campbell, 2004, 2005; Scott, 2001, 2008). The situated, socialized and bounded individuals that give life to a resource governance organization, both its leaders and low-rank members, interpret formal regulation and overall policy mandates (Orihuela et al., 2021). In other words, bricolage begins in the resource governance organization itself, rather than in the community (Funder and Marani, 2015). But the translation of rules and goals into practice is not solely dependent on organizational culture and the wider social context. Interplaying knowledge systems and organizational resources will also condition how, say, a formal rule devised in a conservation area master plan transforms into a tangible, less ambitious everyday practice of monitoring a subset of written 'conservation targets', subject to effective local organizational capacities, the spatial reach of conservationist state and society, and ecological conditions (Orihuela and Mendieta, 2021).

We delve now into the realm of informal rules, referring to the unwritten norms and habitual behaviours as understood in the literature. During fieldwork, it becomes evident that cultural guidelines of conduct are not always the predominant type of informal rules at play in Madre de Dios. In fact, informal and illegal economies wield significant influence vis-à-vis the state, not only in resource-rich peripheral regions, borders and ports of developing countries, but also in their metropolises (Centeno and Portes, 2006; Dewey and Thomas, 2022).

Illicit economies, well-connected to international trade, can exert substantial influence on the effectively operating law, habits and patron-type rules. The case of Peru shows that illegal extractive activities (e.g., minerals, timber, fauna species) and plantation economies (e.g., coca, palm oil, cacao) in the Amazon can have well-established connections and even representation within formal political and legal systems. In the past two decades, regional presidents and congressmen elected in Madre de Dios are local leaders linked to illegal mining and timber (SPDA, 2023a). In a country where 90% of businesses operate informally (OECD, 2016), the intertwining of legal and illegal political-economic systems comes as no surprise (Dammert and Sarmiento, 2019; Dargent, Feldmann and Luna, 2017). If we go back to the fundamentals, resource governance is effectively shaped, perhaps even dictated, by major resource players in the political-economic structure.

Researching resource governance rules and practices in the Andean Amazon

Demonstrating and arguing the influence of a particular set of rules on human behaviour are crucial, rather than making assumptions over it. Drawing from Latour (2005), the social realm can be endlessly reassembled, though perhaps not so easily when natural resource economic power structures are so heavy and stable that no alternative governance has a chance to rise. In specific historical and ecological circumstances, *rules that matter* must first be identified; their functioning and evolution, mapped; their causes, traced; and their behavioural significance, identified and illustrated. Historical and ethnographic methods serve as valuable research tools for this casuistic approach in social science.

How can we trace effective 'structuring constraints' of natural resource use? By focusing on what is practiced rather than what is written. Abers and Keck (2013: 3) define institutions as 'commonly accepted ways of doing things', highlighting the role of actors and organizations in activating new resource governance institutions. In complex, multi-layered, multi-actor,

> entangled institutional environments, where actors must work in multiple arenas to get things done, they often discover that neither the passage of new laws nor the approval of policy designs automatically reconfigures decision making. To influence how other actors behave, they must first develop practical authority.
>
> (Abers and Keck, 2013: 6)

These authors put the focus on behind-the-scenes civil servants and civil society activists, instead of charismatic leaders of policy and institutional change: the soldiers, not the generals, of new resource governance. To matter, new rules and policies need committed activists, individuals who broker the transformation of written rules into practical implementation.

Qualitative methods are better suited for researching the evolution of institutions. In the case of the written rules, my research strategy involved identifying key actors and their epistemic/business/social networks. First, I used primary and secondary sources, followed by interviewing every actor and observer I could identify. This process was less challenging than anticipated because a good part of the protected area-makers and promoters either lived in Lima or were reachable by phone or Skype at the time. Networks were multiple and crosscutting: forestry bureaucrats at the Ministry of Agriculture, La Molina Agrarian University professors and graduates, eco-lodge entrepreneurs, conservation and climate scientists, lay nature lovers, anthropologists, farmers and Indigenous people's groups, with a number of individuals belonging to more than one network.

My first Tambopata paper (Orihuela, 2017) faced challenges due to its limited engagement with local, especially Indigenous, actors, highlighting the need for more extensive fieldwork to establish trust. Particularly, attempts to reach the leadership of the local Indigenous federation FENAMAD went unanswered. My interviews were thus concentrated on only a few Indigenous community members associated with the protected area. Nonetheless, the limited sample of local interviews revealed the heterogeneity of views and interests, especially within Indigenous peoples. This diversity was evident across legally-recognized Indigenous communities and within *Infierno* (Hell), a native community in the buffer zone of the proposed protected area that hosted the first eco-lodge co-owned by white Peruvian entrepreneurs and an Indigenous community (Orihuela, 2017; Stronza, 2000).

As for unwritten rules, we found an opportunity for observation within the institutional arrangement for biological conservation management, under the purview of a non-governmental organization since 2008. We were happy to learn of an intern program for university students, which provided a means to document, and partake in, everyday conservation practices in Tambopata. In addition to participatory observation, primarily based on practicing biological monitoring in two contiguous protected areas for a three-month period,[2] interviewing as many park guards as possible would unveil informal networks of conservation governance and competing assemblages. Fieldwork revealed that not only was there gold mining in the buffer zone, as shown in news sources, but there was also alluvial mining within the protected areas (Orihuela and Mendieta, 2021).

Our studies showed the interconnectedness of the legal and illegal economic webs within the Tambopata and Bahuaja Sonene protected areas. While Tambopata's formal assemblage emphasizes wildlife-watching and science production, there is also a significant presence of gold mining. Interviewees highlighted that the Kotsimba Harakmbut, for instance, work as tour operators in the first web and as miners in the second. In turn, the Ese Eja of Palma Real sometimes travel south to the gold assemblage and become laborers, while others turn to gold mining entrepreneurs in their own river basins. Moreover, some biodiversity monitors and park guards are former gold miners, and there are also forest guards who have nut concessions in the reserve. This complexity underscores that the resource conservation economy is just one facet of the multifaceted economic landscape (Orihuela and Mendieta, 2021).

A final word on the agency of nature. ANT ascribes agency to both natural and social actors, or 'actants'. While the 'actor network' is composed of human and nonhuman elements, the agents of change are those redefining relationships and meanings. Then, what are the non-human actors influencing institutional change in the Tambopata conservation areas? The agency of nature is most evident in the charismatic wildlife watching and science production network, what we call the conservation assemblage. Spatially, this circuit expands around the Sandoval, La Torre and Malinowsky vigilance posts, extended around lakes and sodium-rich hills in the centre of the Tambopata National Reserve. Ecosystem dynamics and the acting of its nonhuman elements make the assemblage feasible. Thus, the visit of colourful birds as well as the soil quality and rain patterns allowing animal footprints to persist for long periods is not uniform across the forest. The original protected area project envisioned a self-sustainable eco-lodge and conservation economy, but nature actors have contributed to redefine actual rules and practices of forest governance (Orihuela, 2017; Orihuela and Mendieta, 2021).

Conclusions

Mainstream institutionalism falls short when inquiring how institutions evolve and matter in the governance of Amazonian forests. First, written rules matter, but not as much as practices, particularly in the world economy's resource periphery contexts. The creation of new written rules for resource governance is a result of a political-economic process, shaped by expert knowledge and transnational power systems. However, designing a paper park says little about how a paper park actually works. Second, effective resource governance is uncertain, multifaceted and fluid. Practices of resource regulation, in particular, are likely to vary across space and time because state reach and embedding socio-ecological systems are not uniform. Frameworks for resource governance inquiry must showcase the interplay of context, power structures and situated agency in customizing new forest governmental practices.

Given the contested and evolving nature of the conservationist state and its written rules, adopting assemblage lenses enhances the study of institutions in relation to resource governance. Whether focusing strictly on formal rules enforcement or more broadly considering new ways of doing things, natural resource governance requires constant forging, reassembling, remaking. The emphasis on practices and the effective translation of formal rules and policy mandates into governmental action necessitates tracing not only the actors of new resource governance but also the organizations and networks that shape them and that they eventually contribute to transform.

By asking us to observe practices (over rules) and change (instead of equilibrium), an assemblage approach is ideal for guiding open-ended empirical research on institutions and natural resource governance. The insight of thinking the social in terms of associations, old and new, acknowledging 'the ephemeral, the indefinite and the irregular', empowers the critical institutionalist eye. With the overall goal of assuming less and explaining more of the remaking of the social environment, ethnographic methods are best suited to study attempted governmental practices regarding nature and to ask how they eventually shape economic conduct. In researching resource governance rules and practices in the Peruvian Amazon, we conclude that protected areas are contested, adaptive and malleable projects under construction. The formation of the Tambopata protected area shows new resource governance progresses, yet its fragility and spatially-diverse influence becomes evident. There is no visible 'green Leviathan', no all-powerful 'anti-politics machine' of conservation or sustainable development. The conservationist state, still effectively weak and in the process of construction, interacts with powerful organized legal/illegal resource economies. The paper park envisioned by new forest governance champions inevitably becomes something multifaceted, contingent on various structural conditions and actors. One forest, multiple attempted assemblages in exchange and dispute.

Notes

1 By referring to "peripheral countries", I am following Raúl Prebisch (1950)s analysis of Latin American economic development, which states that, in the world economic system, peripheral countries provide raw materials to the industrialized core. An updated formulation would have to add global environmental services from forest-abundant developing countries to the contribution of the periphery, or Global South, to the centre, or Global North.

2 Participatory observation took place around 13 "control and vigilance posts" across the protected areas, 11 in the Tambopata Natural Reserve and 2 in the contiguous Bahuaja Sonene National Park. Regularly visited surroundings of protected area posts consist of walking paths, made for biodiversity data collection, along the Tambopata, Madre de Dios, La Torre, Heath, Malinowsky and Inambari rivers.

References

Abers, R. and Keck, M.E. (2013) *Practical Authority: Agency and Institutional Change in Brazilian Water Politics*, Oxford: Oxford University Press.

Acuña, R.M. (2015) The politics of extractive governance: Indigenous peoples and socio-environmental conflicts, *The Extractive Industries and Society*, 2(1): 85–92. https://doi.org/10.1016/j.exis.2014.11.007

Álvarez, A., Alca, J., Galvin, M. and García, A. (2008) The difficult invention of participation in the Amarakaeri Communal Reserve, Peru, in Galvin, M. and Haller, T. (eds.) *People, Protected Areas and Global Change: Participatory Conservation in Latin America, Africa, Asia and Europe*, Perspectives of the Swiss National Centre of Competence in Research (NCCR) North-South, Volume 3. Bern: University of Bern, pp. 111–144.

Alvarez, N.L. and Naughton-Treves, L. (2003) Linking national agrarian policy to deforestation in the Peruvian Amazon: A case study of Tambopata, 1986–1997, *AMBIO: A Journal of the Human Environment*, 32(4): 269–274. https://doi.org/10.1579/0044-7447-32.4.269

Álvarez-Berríos, N.L., L'Roe, J. and Naughton-Treves, L. (2021) Does formalizing artisanal gold mining mitigate environmental impacts? Deforestation evidence from the Peruvian Amazon, *Environmental Research Letters*, 16(6): 064052. https://doi.org/10.1088/1748-9326/abede9

Asiyanbi, A.P., Ogar, E. and Akintoye, O.A. (2019) Complexities and surprises in local resistance to neoliberal conservation: Multiple environmentalities, technologies of the self and the poststructural geography of local engagement with REDD+, *Political Geography*, 69: 128–138. https://doi.org/10.1016/j.polgeo.2018.12.008.

Bevir, M. (1999) Foucault, power and institutions, *Political Studies*, 47: 345–359. https://doi.org/10.1111/1467-9248.00204

Bevir, M. (2010) Rethinking governmentality: Towards genealogies of governance, *European Journal of Social Theory*, 13(4): 423–441. https://doi.org/10.1177/1368431010382758

Blackman, A., Corral, L., Lima, E.S. and Asner, G.P. (2017) Titling indigenous communities protects forests in the Peruvian Amazon, *Proceedings of the National Academy of Sciences*, 114(16): 4123–4128. https://doi.org/10.1073/pnas.1603290114

Brinks, D.M., Levitsky, S. and Murillo, M.V. (2019) *Understanding Institutional Weakness: Power and Design in Latin American Institutions*, Cambridge: Cambridge University Press. https://doi.org/10.1017/9781108772211

Bromley, P. and Powell, W.W. (2012) From smoke and mirrors to walking the talk: Decoupling in the contemporary world, *Academy of Management Annals*, 6(1): 483–530. https://doi.org/10.5465/19416520.2012.684462

Campbell, J.L. (2004) *Institutional Change and Globalization*, Princeton: Princeton University Press.

Campbell, J.L. (2005) Where do we stand? Common mechanisms in organizations and social movements, in Davis, G.F., McAdam, D. and Scott, W.R. (eds.) *Social Movements and Organization Theory*, New York: Cambridge University Press, pp. 41–68. https://doi.org/10.1017/CBO9780511791000.004

Centeno, M.A. and Portes, A. (2006) The informal economy in the shadow of the state, in Fernández-Kelly, P. and Shefner, J. (eds.) *Out of the Shadows: Political Action and the Informal Economy in Latin America*, Pennsylvania: Penn State University Press.

Chavarría, M.C., Rummenhöller, K. and Moore, T. (eds.) (2020) *Madre de Dios: Refugio de Pueblos Originarios*, Lima: USAID.

Choi, M.-A. (2020) Multiple environmental subjects: Governmentalities of ecotourism development in Jeungdo, South Korea, *Geoforum*, 110: 77–86. https://doi.org/10.1016/j.geoforum.2020.01.011

Cleaver, F. (2002) Reinventing institutions: Bricolage and the social embeddedness of natural resource management, *The European Journal of Development Research*, 14(2): 11–30. https://doi.org/10.1080/714000425

Cleaver, F. (2012) *Development through Bricolage: Rethinking Institutions for Natural Resource Management*, London: Routledge. https://doi.org/10.4324/9781315094915

Cleaver, F. and Whaley, L. (2018) Understanding process, power, and meaning in adaptive governance: A critical institutional reading, *Ecology and Society*, 23(2): 49. https://www.jstor.org/stable/26799116

Cortes-Vazquez, J.A. and Ruiz-Ballesteros, E. (2018) Practising nature: A phenomenological rethinking of environmentality in natural protected areas in Ecuador and Spain, *Conservation and Society*, 16(3): 232–242. https://doi.org/10.4103/cs.cs_16_158

Dammert, J.L. (2015) *Hacia una ecología política de la palma aceitera en el Perú*, Lima: Oxfam.

Dammert, J.L. (2017) *Acaparamiento de tierras en la Amazonia peruana, el caso de Tamshiyacu*, Lima: Wildlife Conservation Society.

Dammert, L. and Sarmiento, K. (2019) Corruption, organized crime, and regional governments in Peru, in Rotberg, R.I. (ed.) *Corruption in Latin America: How Politicians and Corporations Steal from Citizens*, Cham: Springer, pp. 179–204. https://doi.org/10.1007/978-3-319-94057-1_8

Damonte, G.H. (2018) Mining Formalization at the Margins of the State: Small-scale Miners and State Governance in the Peruvian Amazon, *Development and Change*, 49(5): 1314–1335. https://doi.org/10.1111/dech.12414

Dargent, E., Feldmann, A.E. and Luna, J.P. (2017) Greater state capacity, lesser stateness: Lessons from the Peruvian commodity boom, *Politics and Society*, 45(1): 3–34. https://doi.org/10.1177/0032329216683164

Dean, M. (2009) *Governmentality: Power and Rule in Modern Society*, 2nd edition, London: Sage.

Dewey, M. and Thomas, K. (2022) Futurity beyond the state: Illegal markets and imagined futures in Latin America, *Latin American Politics and Society*, 64(4): 1–23. https://doi.org/10.1017/lap.2022.24

Dourojeanni, M.J. (2009) *Crónica Forestal del Perú*, Lima: UNALM and UNMSM.

Dourojeanni, M.J. (2018) *Áreas Naturales Protegidas del Perú: El Comienzo*, Lima: UNGV.

Drinot, P. (2011) The meaning of Alan García: Sovereignty and governmentality in neoliberal Peru, *Journal of Latin American Cultural Studies*, 20(2): 179–195. https://doi.org/10.1080/13569325.2011.588514

Evans, P. (2004) Development as institutional change: The pitfalls of monocropping and the potentials of deliberation, *Studies in Comparative International Development*, 38: 30–52. https://doi.org/10.1007/BF02686327

Ferguson, J. (1994) *The Anti-Politics Machine: "Development," Depoliticization, and Bureaucratic Power in Lesotho*, Minneapolis: University of Minnesota Press.

Fletcher, R. (2010) Neoliberal environmentality: Towards a poststructuralist political ecology of the conservation debate, *Conservation and Society*, 8(3): 171–181. https://doi.org/10.4103/0972–4923.73806

Foucault, M. (2008) *The Birth of Biopolitics*, New York: Palgrave Macmillan.

Funder, M. and Marani, M. (2015) Local bureaucrats as bricoleurs. The everyday implementation practices of county environment officers in rural Kenya, *International Journal of the Commons*, 9(1): 87–106. https://doi.org/10.18352/ijc.526

Furumo, P.R. and Aide, T.M. (2017) Characterizing commercial oil palm expansion in Latin America: Land use change and trade, *Environmental Research Letters*, 12(2): 024008. https://doi.org/10.1088/1748-9326/aa5892

Hall, K., Cleaver, F., Franks, T. and Maganga, F. (2014) Capturing critical institutionalism: A synthesis of key themes and debates, *The European Journal of Development Research*, 26(1): 71-86. https://doi.org/10.1057/ejdr.2013.48

Hochstetler, K. and Keck, M.E. (2007) *Greening Brazil: Environmental Activism in State and Society*, Durham, NC: Duke University Press.

Hughes, N. (2010) Indigenous protest in Peru: The 'Orchard Dog' bites back, *Social Movement Studies*, 9(1): 85–90. https://doi.org/10.1080/14742830903442550

Infobae. (2024) Ley forestal: Congreso omite reconsideraciones y ordena promulgar ley que permite la deforestación ilegal. Available at https://www.infobae.com/peru/2024/01/10/ley-forestal-congreso-omite-reconsideraciones-y-envia-al-gobierno-autografa-que-permite-la-deforestacion-ilegal/. Accessed on January 24, 2024.

Jacopo, C. (2022) The nexus between corruption and money laundering: Deconstructing the Toledo-Odebrecht network in Peru, *Trends in Organized Crime*, online first. https://doi.org/10.1007/s12117-021-09439-6

Latour, B. (2005) *Reassembling the Social: An Introduction to Actor-Network-Theory*, Oxford: Oxford University Press.

Law, J. (2004) *After Method: Mess in Social Science Research*, Abingdon: Routledge.

Law, J. and Mol, A. (2008) The actor-enacted: Cumbrian sheep in 2001, in Knappett, C. and Malafouris, L. (eds.) *Material Agency: Towards a Non-Anthropocentric Approach*, Boston, MA: Springer, pp. 57–78. https://doi.org/10.1007/978-0-387-74711-8_4

Levitsky, S. and Murillo, M.V. (2009) Variation in institutional strength, *Annual Review of Political Science*, 12: 115–133. https://doi.org/10.1146/annurev.polisci.11.091106.121756

Li, T.M. (2007) *The Will to Improve: Governmentality, Development, and the Practice of Politics*, Durham, NC: Duke University Press.

Li, T.M. (2014) *Land's End: Capitalist Relations on an Indigenous Frontier*, Durham, NC: Duke University Press.

Li, T.M. and Semedi, P. (2021) *Plantation Life: Corporate Occupation in Indonesia's Oil Palm Zone*, Durham, NC: Duke University Press.

Mahoney, J. and Thelen, K. (2010) A theory of gradual institutional change, in Mahoney, J. and Thelen, K. (eds.) *Explaining Institutional Change: Ambiguity, Agency, and Power*, Cambridge: Cambridge University Press, pp. 1–37.

Manrique, H. and Orihuela, J.C. (2024) Slow environmental justice: The Cuninico oil spill and the legal struggle against oil pollution in Peruvian Amazonia, *Environmental Politics*, online first. https://doi.org/10.1080/09644016.2024.2322382

Massi, E. and Nem Singh, J. (2018) Industrial policy and state-making: Brazil's attempt at oil-based industrial development, *Third World Quarterly*, 39(6): 1133–1150. https://doi.org/10.1080/01436597.2018.1455144

Moore, T. and Chicchón, A. (1997) Tambopata-Candamo, Peru: Eco-regional strategy for Sustainable development, in Ornat, A.L. (ed.) *Strategies for Sustainability: Latin America*, Sterling, VA: Earthscan, pp. 146–154. https://doi.org/10.4324/9781315066134

North, D.C. (1990) *Institutions, Institutional Change and Economic Performance*, Cambridge: Cambridge University Press. https://doi.org/10.1017/CBO9780511808678

OECD. (2016) *Multi-dimensional Review of Peru: Volume 2. In-depth Analysis and Recommendations*, OECD Development Pathways, Paris: OECD Publishing. https://doi.org/10.1787/9789264264670-en

Orihuela, J.C. (2014) Converging divergence: The diffusion of the green state in Latin America, *Studies in Comparative International Development*, 49: 242–265. https://doi.org/10.1007/s12116-013-9147-6

Orihuela, J.C. (2017) Assembling participatory Tambopata: Environmentality entrepreneurs and the political economy of nature, *Forest Policy and Economics*, 80: 52–62. https://doi.org/10.1016/j.forpol.2017.03.010

Orihuela, J.C. (2018) Institutions and place: Bringing context back into the study of the resource curse, *Journal of Institutional Economics*, 14(1): 157–180. https://doi.org/10.1017/S1744137417000236

Orihuela, J.C. (2020) Embedded countermovements: The forging of protected areas and native communities in the Peruvian Amazon, *New Political Economy*, 25(1): 140–155. https://doi.org/10.1080/13563467.2019.1570101

Orihuela, J.C. and Mendieta, A. (2021) One and three forests: Understanding institutional diversity in Amazonian protected areas, *Forest Policy and Economics*, 133: 102620. https://doi.org/10.1016/j.forpol.2021.102620

Orihuela, J.C., Mendieta, A., Pérez, C. and Ramírez, T.L. (2021) From paper institutions to bureaucratic autonomy: Institutional change as a resource curse remedy, *World Development*, 143: 105463. https://doi.org/10.1016/j.worlddev.2021.105463

Palacios, D.P. and Sarmiento, J.P. (2021) The meeting of multiple governmentalities and technologies of participation in protected areas: The case of the Amarakaeri Communal Reserve (Peruvian Amazon), *International Forestry Review*, 23(1): 90–102. https://doi.org/10.1505/146554821833466095

RAISG (2020) *Amazonía bajo Presión*. Avaliable at https://atlas2020.amazonia-socioambiental.org/en Accessed on January 4, 2024.

Santos-Granero, F. and Barclay, F. (2011) Bundles, stampers, and flying gringos: Native perceptions of capitalist violence in Peruvian Amazonia, *The Journal of Latin American and Caribbean Anthropology*, 16(1): 143–167. https://doi.org/10.1111/j.1935-4940.2011.01128.x

Scott, W.R. (2001) *Institutions and Organizations*, 2nd edition, Thousand Oaks, CA: Sage.

Scott, W.R. (2008) Approaching adulthood: The maturing of institutional theory, *Theory and Society*, 37(5): 427–442. https://doi.org/10.1007/s11186-008-9067-z

Smith, R.C. and Pinedo, D. (eds.) (2002) *El cuidado de los bienes comunes: gobierno y manejo de los lagos y bosques en la Amazonía*, Lima: Instituto de Estudios Peruanos.

Smith, R.C. and Pinedo, D. (2003) Comunidades y áreas naturales protegidas en la Amazonía peruana, *Debate agrario*, 36: 15–37.

SPDA. (2023a) ¿Cómo Madre de Dios se convirtió en un lugar tan inseguro para los defensores ambientales? Available at https://www.actualidadambiental.pe/madre-de-dios-se-convirtio-en-un-lugar-tan-inseguro-defensores-ambientales/. Accessed on January 4, 2024.

SPDA. (2023b) La SPDA rechaza modificación de Ley Forestal y de Fauna Silvestre que legalizaría la deforestación. Available at https://spda.org.pe/pronunciamiento-la-spda-rechaza-modificacion-de-ley-forestal-y-de-fauna-silvestre-que-legalizaria-la-deforestacion/. Accessed on January 4, 2024.

Stronza, A.L. (2000) *Because It Is Ours: Community-Based Ecoturism in the Peruvian Amazon*, PhD Thesis, Gainesville: University of Florida.

Tsing, A.L. (2005) *Friction: An Ethnography of Global Connection*, Princeton, NJ: Princeton University Press.

Vijay, V., Reid, C.D., Finer, M., Jenkins, C.N. and Pimm, S.L. (2018) Deforestation risks posed by oil palm expansion in the Peruvian Amazon, *Environmental Research Letters*, 13(11): 114010. https://doi.org/10.1088/1748-9326/aae540

Winkel, G. (2012) Foucault in the forests – A review of the use of "Foucauldian" concepts in forest policy analysis, *Forest Policy and Economics*, 16: 81–92. https://doi.org/10.1016/j.forpol.2010.11.009

11 Navigating rules of 'the range' beyond the commons

Focusing on relationality and process in China and Kenya

Palden Tsering and Ryan R. Unks

A shift: formation to functioning

Max Weber was the forerunner of institutional focus and study, emphasising the establishment and the interrelationship of the religious revolution and capitalist institutions in 16th-century Europe. Weber saw that the fundamental religious revolutionary moral elements, such as trust, hard work and sacrifice, were significant for capitalistic socio-economic development. As a result, the link between culture and institution was highlighted, which had a lasting impact on sociological approaches. However, many schools of thought on institutions have emerged since the 1970s, including new institutionalism and sociological institutionalism. Although the term 'institution' has gained burgeoning attention in the scientific community, there have been many interpretations and analytical framings over time. Numerous researchers use institutions to refer to 'rules' and 'norms' (North, 1987). According to Elinor Ostrom, rules are 'prescriptions that are commonly known and used by a set of participants to order repetitive, interdependent relationships, and they are potentially subject to change' (Ostrom, 1986: 5–6). Such rules, as Agrawal and Perrin explain, are often 'humanly created formal and informal mechanisms that shape social and individual expectations, interactions, and behaviour' (2009: 1).

The school of historical institutionalism defines institutions as the formal and informal routines, norms and conventions embedded in organizations (Hall and Taylor, 1996). Institutions are often associated with organizations and the rules or conventions promulgated by formal organizations (Hall and Taylor, 1996). Historical institutionalists have paid close attention to how institutions distribute power unevenly across social groups, using calculus and cultural approaches (Hall and Taylor, 1996; Ostrom, 1986). Sociological institutionalists argue that culture is essential for explaining institutions, as it is central in influencing practice and decisions. As Hall and Taylor state, 'institutional forms and procedures used by modern organizations were not adopted simply because they were most efficient... even the most seemingly bureaucratic practices, have to be explained in cultural terms' (1996: 14).

Cleaver divides these institutional approaches into two schools of thought: Mainstream Institutionalism and Critical Institutionalism (Cleaver, 2012;

DOI: 10.4324/9781032624679-11

see also Chapter 1 of this volume), where mainstream approaches are comparatively instrumental and linear, and critical approaches are more complex, nonlinear and 'fuzzy'. From this perspective, in mainstream institutional theory, the nature of institutions is 'formal and functional; they are crafted to address a specific dilemma of resource management; effectiveness is equated with clarity of purpose, transparency, public accountability and regularity of operation' (Cleaver, 2012: 13). Mainstream approaches 'provide information and assurance about the behaviour of others, to offer incentives to behave for the collective good, and to monitor and sanction opportunistic behaviour' (Cleaver, 2012: 8). In contrast, critical institutionalism aims to explore how various 'institutions dynamically mediate relationships between people, natural resources, and society' (Cleaver and de Koning, 2015: 1). This approach highlights the complexity of institutions, and the incorporation of traditional and modern, informal and formal arrangements, and are 'often multi-purpose, intermittent and semi-opaque in operation' (Cleaver, 2012: 13; see also Cleaver, 2001; Lund, 2006). This emphasizes a mix of economic, emotional, moral, cosmological and social rationalities informed by differing logics and world-views (Bennett et al., 2018; Cleaver and de Koning, 2015; Hall et al., 2014; Whaley, 2018). Furthermore, mainstream institutional approaches often underestimate and overlook the complex and dynamic nature of institutional governance in the socio-economic system (Cleaver, 2012).

While various theorists interpret institutions differently, many underline their complexity and embeddedness in day-to-day life. Richard Scott defines institutions as 'socially constructed symbolic systems that, together with associated activities and resources, provide stability and meaning to social life' (2014: 45). Scott categorized institutions into three types, or as he puts it, 'three basic pillars,' which are: regulative, normative and cultural-cognitive. However, on the ground, these socially constructed symbolic systems rarely work in isolation. By contrast, many formal and informal institutions in rural settings are often in a state of 'messiness' and overlap, and are located in particular cultural, social and political contexts (Scoones, 2015; Waylen, 2014). After all, the formal and informal are constantly interchanging, where formality skirts around informality, and *vice versa*. As discussed in Chapter 1 of this volume, the boundary between informal and formal is blurry, and a dichotomous perspective needs to be questioned.

In pastoral China and Kenya, increased exploration and competition for natural resources, notably land requisition, commercialization and preservation of land, and increasing focus on wildlife conservation, green energy, carbon sequestration and mining, have frequently resulted in pastoralists losing their rights and ability to access rangeland and resources on the ground. Consequentially, informal and formal institutions that regulate access are being subtly reconfigured (Bauer and Nyima, 2011; Gelek, 2002; Gruschke, 2008; Lind, Okenwa and Scoones, 2020; Nori and Scoones, 2019; Tsering, 2023; Unks et al., 2019, 2023; Yeh et al., 2017).

Existing institutional arrangements for rangeland use and access are being altered through time and space by various actors and at multiple levels of social organization, resulting in newly-emerging institutional arrangements for land governance (German et al., 2017a; Tsering, 2023; Unks et al., 2019, 2023). This is also seen elsewhere in processes of shifts in land control (Peluso and Lund, 2011; Rasmussen and Lund, 2018). As Li notes, governmental interventions that are designed to 'improve' the world are 'assembled from diverse elements, such as discourses, institutions, forms of expertise and social groups whose deficiencies need to be corrected, among others' (Li, 2007: 263). In rural China, the assumed necessity to 'civilize,' foster 'progress' and 'develop' the 'backward', 'conservative' western regions (Yeh, O'Brien and Ye et al., 2013) became the People's Republic of China (PRC)'s primary concern after de-collectivization and the shift to a market economy in the 1980s (Fischer, 2008; Gruschke, 2012; Heilmann and Perry, 2011; Roche and Hillman, 2017; Topgyal, 2011; Yeh et al., 2017). Similarly, in Kenya, a long history of interventions in pastoral institutions, based on similarly pejorative assumptions, have attempted to convert pastoral land use to commercial livestock production, to sedentarize populations and to impose regulatory grazing management (Anderson, 2002; Mwangi, 2007; Mwangi and Ostrom, 2009; Okoth-Ogendo, 1986, 1989; Unks, 2022; Waller, 2012).

Rules of the range: assemblage is always in motion

To analyze hybrid property and land access relations in pastoral settings, where rules, roles and relationships between rangelands and their users are in a perpetual state of flux, we expanded on critical approaches to studying institutions to explore the roles of diverse sets of actors that shape these institutions. Utilizing the notion of assemblage as an analytical tool helped us to direct attention away from notions of completed or settled property rights onto ongoing processes of drawing heterogeneous elements together, forging connections between them and sustaining these relationships in the face of uncertainty. Assemblage emphasizes the encounter of different processes and practices, echoing other concepts, such as bricolage, pluralism and negotiations. Cleaver defines institutional bricolage as 'a process in which people... draw on existing social formulae ... to patch or piece together institutions in response to changing situations' (Cleaver, 2012: 45). The concept of bricolage has been used to show how institutions tend to form out of a complex mix of elements (e.g. customary, colonial-state era, post-independence state, non-governmental), are social and relational, and influenced by political, economic, historical and discursive processes (Cleaver, 2012; Cleaver et al., 2013; Hall et al., 2014; Whaley, 2018). Both assemblage and bricolage blur the focus on where different types of power are located, along with debates about structure vs. agency (Cleaver, 2012) and necessitates understanding institutions as relational processes to be able to understand power. They also both similarly move away from a normative focus on 'getting the rules right' to sustain common pool resource use (Johnson, 2004), leading to a more holistic understanding of how local institutions often represent powerful interests and have evolved in relation to wider global phenomena (Cleaver, 2012; Hall et al., 2014; Mosse, 1997). However, assemblage, as we apply it here, adds to the concept of bricolage by drawing our focus to diverse actors and different resource governance practices at different levels of organization.

Assemblage has been interpreted in a variety of ways in recent decades, and scholars employ it in a variety of circumstances. Deleuze and Guattari (taken from Anderson and McFarlane, 2011: 125) describe assemblage as 'a process of 'co-functioning' in which diverse parts join together in a non-homogeneous grouping'. Building on this, science and technology studies used similar conceptualizations of assemblage to account for the hybridity and heterogeneity of elements in actor-network theory (Latour, 2005). In her work, Tania Murray Li describes assemblage as 'the on-going labour of bringing disparate elements together and forging connections between them' (Li, 2007: 263). The notion of assemblage entails the convergence of a diverse array of agents with different objectives who draw on different relationships, technologies and discourses to intervene in social processes to achieve desired outcomes while mitigating undesired ones (Anderson and McFarlane, 2011; Li, 2007). Again, the practice of reassembling described in Li's study resonates with Cleaver's bricolage (2012); she stresses that 'reassembling' refers to the 'grafting on new elements, and reworking old ones', such as deploying new discourses to new ends and transporting the meanings of key terms (Li, 2007: 265). Although the terms 'assemblage' and 'bricolage' have different *foci*, they both stress practices that occur in a nonlinear, messy and changing context. These concepts pose a 'how' question, specifically: how the actions of connecting different materials at hand are arranged under tension, uncertainties and changes in different contexts.

The notion of assemblage is aimed at understanding the whole picture of a given socio-cultural and political economic conjuncture, thus, 'deliberately destabilising current categories and definitions' (Aneesh, 2017: 129; see also Li, 2007, 2014; Sassen, 2008). As Anderson and McFarlane (2011) point out, assemblage should be thought of as a temporary process since new components may enter, relationships between actors can change, disputes and coercions can arise, and alliances can be destroyed (see also Li, 2014). Thus, assemblage entails an epistemic orientation toward understanding how 'relations emerge, take hold, and endure, but may also change or be disrupted' (Li, 2014: 125). This flattens the ontology of institutions that we apply in our work, where we conceptualize their origins working in either top-down ways (e.g. part of a 'governmentality' where diverse powerful actors are brought together through common interests in modifying institutions, Dean, 2010), or as pastoralists assembling institutions in an empowering way. We apply the idea of assemblage as the practice of bringing together an array of agents (pastoralists, religious organizations, local governments and investors, non-governmental actors) and objectives (poverty alleviation, cultural preservation, livelihoods and reciprocal relationships, industrial extraction, biodiversity conservation) to 'intervene in social processes to produce desired outcomes and avert undesired ones' (Li, 2007: 264).

Rangeland governance in the Tibetan-Chinese context is often divided into a binary communal or private use, with the latter being incentivised by the 'grassland contracting policy' from the 1980s. Yet, on the ground, rangeland governance is much more complicated, dynamic, and flexible. Similar examples of changes following waves of privatization, but ongoing collective use, stand out in the Kenyan context (Archambault, 2016; Jeppesen and Hassan, 2022). Here, on the ground, the differences between 'right' and 'ability' help to clarify the 'enforceable claim' and the ability to access (Peluso and Ribot, 2003; 2020), under the rangeland contract policy in pastoral regions (Nyima, 2021). Ownership and use rights had been authorized from the state contract in China, where the contract claims the contractor's use right; however, do contractors have the ability to constrain the use of and access to the contracted rangeland? Similarly, in the Kenyan context, wildlife conservancies have recontoured access regimes where pastoralists retain formal titles over land, but the involvement of transnational actors regulating access in complex ways raises questions about how access and benefits play out in practice (Unks et al., 2023). As Peluso and Ribot (2003; 2020) raise, formal rights such as those that the state guarantees, do not ensure access and the reality of access is fuzzy and messy on the ground.

This notion of hybrid rangeland governance, as practices of assemblage that combine heterogeneous elements, such as rules, relations and roles among resource user groups for *de facto* gain, contributes to the commoning literature's focus on social processes and relations. The term 'commons' or 'common pool resources' refers to something distinct from the property rights systems governing and managing them. According to Ostrom et al. (1999: 19), 'commons refers to regulated resource use, as opposed to open access, which lacks property arrangements'. Conceptual issues with this framing arise when considering how clearly defined property rights are often emphasized as necessary to sustain collective action within bounded common property systems. In pastoral systems, variable ecological conditions, seasonal mobility requirements and other adaptive responses in mobility often make these boundaries 'fuzzy' at best (Fernández-Giménez, 2002; Turner, 2011). Closely related, central design principles found in the work of Ostrom (1990), namely that of a clearly defined group and the ability to exclude outsiders, are thought to be hallmarks of a sustainable commons in the mainstream institutionalism literature. This has led to scholars emphasizing fluidity in practice between conceptually distinct categories of state, common and private land (Robinson, 2019) and emphasizing a balance between tenure security and flexibility (Turner and Schlect, 2019). Different conceptualizations of legally pluralistic property have occurred in parallel to studies of pastoralism that emphasize varying degrees of privatization and exclusion in pastoralist lands (German et al., 2017b) and properties of pastoralist governance that emphasize complex mosaics of control of access based upon the variability and importance of resources (Robinson, 2019; Unks et al., 2023). Synthesizing these pluralistic views of complex property relations that shape resource access, emphasizes variable access, where multiple layers of institutions mediate responses and resource use (Unks et al., 2023). Rather than focusing on institutions as being designed to favour one configuration of property or another, or use of one type of resource or another, focusing on assemblage leads to understanding institutions themselves as inherently political, social and relational entities that are undergoing constant change.

Additionally, 'new commons' and the commoning scholarship have shown the importance of treating commons as having significant social dimensions, as well as addressing the question of how 'individual and collective acts are not always conscious, but rather are relational outcomes of subjectification,[1] individual agency, emotion, and embeddedness within larger political economies' (Nightingale, 2011, 2013: 19; see also, Velicu and García-López, 2018). Commons is not a resource or place, but rather 'a set of more-than-human, contingent relations-in-the-making that result in collective practices of production, exchange and living with the world' (Nightingale, 2019: 18). Commons is the social form[2] of matter that is constructed through practices of commoning,[3] the core of what makes commons what they are (Euler, 2018), and a set of social practices and performances that foster new relations and subjectivities, but these relations are always contingent, ambivalent outcomes of the exercise of power (Nightingale 2019: 16). Rather than seeking to cement property rights, relations of sharing and collective practices are the backbone of durable commoning efforts. Thus, while previous analyses provided insight, especially amidst calls for privatization (e.g. see Ostrom, 1990), we situate ourselves in a perspective that moves institutional analysis away from focusing on the lack of 'fit' of institutions, and shifts in land tenure from collective management toward privatized forms, despite the need for mobility to maintain livestock in semi-arid environments (Lesorogol, 2008; Mwangi, 2007).

The commons are not just resources to be governed through new 'optimal' institutional arrangements, but a whole suite of personal, political, ethical and relational practices around the production of value and solidarity that emerge in the new commons, from the internet to collective approaches to finance or urban living. Commons do not simply exist; they depend on the relationships and interactions that people have with each other and with the goods in question (Helfrich, 2012; Helfrich and Heinrich-Böll-Stiftung, 2009). Thus, commons are part of a web of relationships, both concrete matter and a process in motion, all in one (Mies and Bennholdt-Thomsen, 2001). Critical property studies have expanded the definition of 'property' from a focus on institutions and rights, to conceptualize property as a relationship that links social actors with objects of value (Sikor, 2008; von Benda-Beckmann, von Benda-Beckmann and Eckert, 2009). Furthermore, property rights are meaningless if they cannot be exercized, and many contexts have overlapping and competing relationships that govern resources (Côte and Korf, 2018; Lund, 2008; Peluso, 1996; Sikor and Lund, 2009). Property in this sense is conceptualized as power-laden social relationships that emerge around land or resources, rather than an object or rights (Sikor and Lund, 2009; von Benda-Beckmann, von Benda-Beckmann and Eckert, 2009). Similarly, while bundles of 'rights' such as access, withdrawal, management, exclusion and alienation are often emphasized (Schlager and Ostrom, 1992) critical approaches tend to move beyond rights, for example, building on Ribot and Peluso's work (2003), to consider the structural and relational, social, political and economic 'webs and bundles of powers' that mediate access to land. These approaches offer a more nuanced understanding of how institutions are historically contingent and intertwined with authority and power (Cleaver and de Koning, 2015; Sikor and Lund, 2009).

Navigating rules of the range in China and Kenya

Probing hybrid institutional practices on the ground, and how such hybridity is constructed through assemblage, requires carefully designed research methodologies. We relied heavily on ethnographic methodologies including, semi-structured in-depth interviews, key-informant interviews, participant observation, informal interviews/back stage talks/casual chat and photovoice. We show the efficacy of these ethnographic methods for exploring informal, pluralistic aspects of how institutional systems assemble and evolve, the relationship between local governance and 'top-down' interventions, and asymmetries in social, political and economic power (Figure 11.1).

Two ethnographic case studies were collected from China and Kenya, one case illustrating how pastoralists assemble rules at hand to maintain and secure their rights to access, another how, as pastoralists have sought security of rights over land, the process of assemblage has been dominated by NGOs and state actors. In what we present here, we describe practices of assemblage in two cases that were brought on by uncertainties that were opened by shifts in land governance policies from above. These two cases demonstrate different dynamics of how, through assemblages of diverse actors, institutional arrangements were reconfigured. Through immersive fieldwork, we were able to develop understandings of how different rules and norms were applied in practice, to assess their origins and to assess the perspectives of different actors on why institutional changes were happening and how these changes were impacting practices of mobility and access. These multiple ethnographic methods enabled us to understand informal, pluralistic aspects of how institutional systems assemble and evolve, the relationship between local governance and 'top-down' interventions, and asymmetries in social, political and economic power. These methods also provided the basis of in-depth analysis of institutional arrangements, with particular attention to who is involved in land

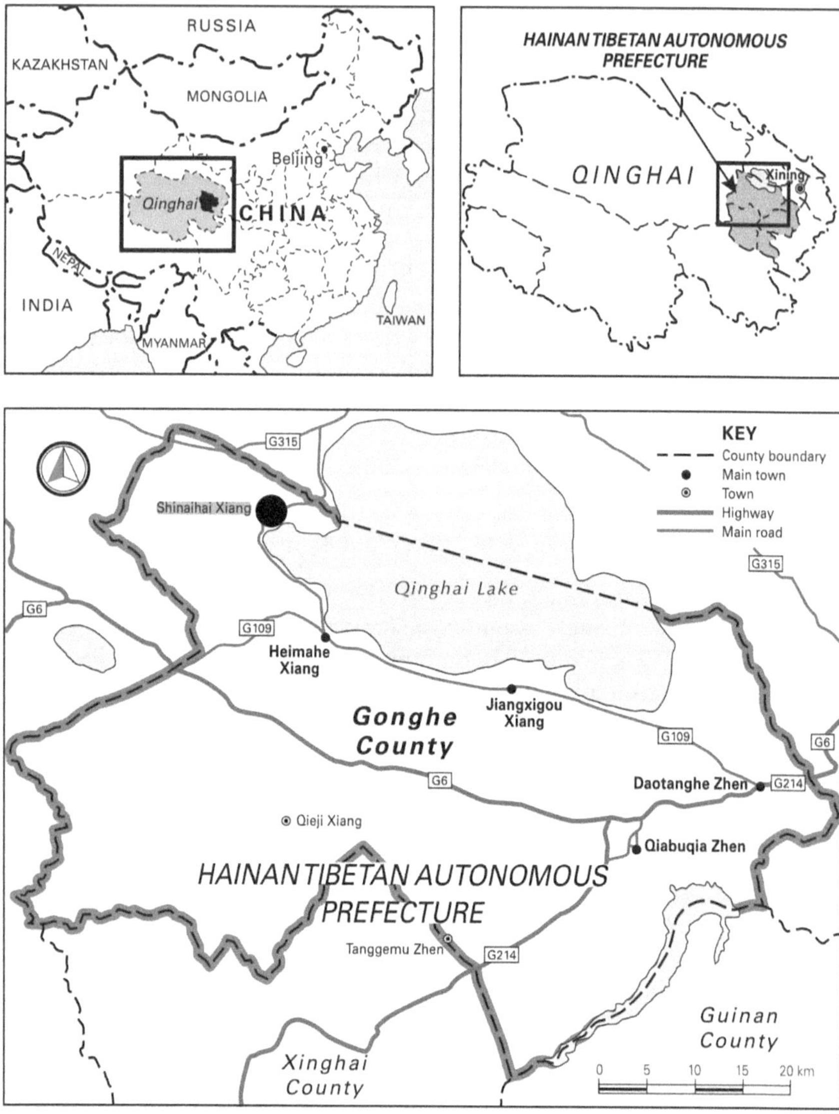

Figure 11.1 Saga in Amdo Tibet, China.

governance, relationships between different actors, and power gradients and politics of inclusion/exclusion. We then explored how formal and informal institutions and their interrelations shape processes of institutional change in both settings, and how access by different groups affects inclusion and exclusion in the process.

In each case, a range of factors were coded for analysis, including actors (consisting of pastoralists in different villages with different levels of influence,

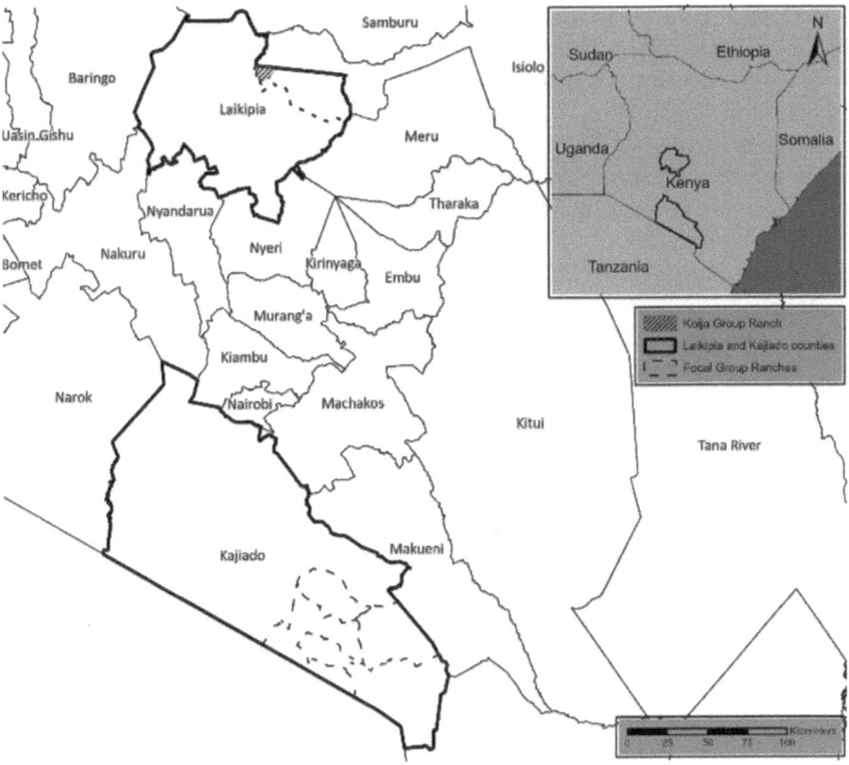

Figure 11.2 Laikipia and Kajiado Counties, Kenya.

investors, workers from resource bureaus, NGOs and government officials), institutions (such as the laws, regulations, customary practices), relationships (consisting of the village, pastoralist with state and local authorities) and interactions (such as compliance, resistance, collusion and cooperation).

China case study: policies from above, solutions from below

The Rangeland Ecological Protection Subsidy and Reward Mechanism (REPSRM) (*Caoyuan shengtai baohu buzhu jianglijizi*) programme,[4] as a priority rangeland state policy for rangeland conservation, is implemented by the provincial government downwards through the county and township governments. The REPSRM programme aims to protect rangeland in rural pastoral regions by banning grazing and limiting livestock numbers.

This research was conducted from 2019 to 2022 at Saga village, located in the north-western part of the lake Kokonor region. With an average of

3,200 m altitude, Saga has 366 households and a total of 1,245 people (Hainan TAP, 2019). The pastoralists graze 65,434 head of livestock, and sheep account for nearly 87% of the total livestock numbers in Saga. Local public authorities, such as village leaders, had ultimate authority over grazing ban sites within the village. While rangeland degradation and the need for conservation were prominent in state narratives in Saga, pastoralists established the zones to maintain the collectively-valued utilization of the pasturelands. As this shows, the REPSRM programme's design is not grounded in research and is not founded on an understanding of the established roles, rules and relationships in pastoralist settings, such as Saga. Summer pasture is used collectively because pastoralists understand the ecological conditions of the pasture. Many interlocutors concluded that summer pasture is comparatively smaller than winter pasture, unsuitable for lengthy grazing due to weather conditions and the scarce vegetation; thus, leaving summer pasture under communal use is intended to improve its utilization and sustainability.

In the case of the grazing ban and balanced zones in Saga, the state and the justice bureau's directives and regulations were reinvented through local practices. Due to the length of utilization, distance from the township centre, and size of the land, pastoralists established the summer pasture as the grazing ban zone, and the privatized winter pasture as the balanced zone. In addition, while the REPSRM programme aims to constrain rangeland use and access in rural pastoralist regions by prohibiting grazing and limiting livestock numbers, the institutional arrangements governing land use and access are constructed through social and political negotiations. As a result, new rules incorporating elements of the state and local society emerge, but these rules remain contested depending on the context, and as circumstances change. As the case revealed, the summer pasture was chosen as the grazing ban zone to evade government inspection, whereas pastoralists benefited from the location, distance and road condition between the township and the alpine meadow that dominated the summer pasture. Thus, the summer pasture's location facilitated pastoralists' ability to maintain land use. Meanwhile, despite the fact that the township government is aware of the uninterrupted use of the summer pasture, they choose to remain silent about it.

Rangeland governance is always in motion in Saga – meaning it is continuously evolving, not only in respect to new top-down interventions and rules introduced by the state but also through emerging bottom-up practices. Pastoralists from Saga not only have a complete understanding of the rules and regulations from the government on rangeland utilization, but also convert these fiats to suit their needs. As the township government confirmed the village grazing ban zone in accordance with the grazing areas proposed by pastoralists, the question of whether the proposed pasture complies with the requirements of the REPSRM programme was deliberately neglected. In other words, rules are not fixed or settled; they are challenged and reinterpreted to suit local realities and needs. As one of the village leaders explained when asked about the contradiction of herding livestock within the grazing ban zone while receiving the subsidy from the state, he explained:

> The life of the pastoralists depends on the livestock and rangeland. We always do what we can to make the pastoralists benefit from the pasture. The state wants to solve the problem of rangeland degradation by compensating us and making us reduce grazing and livestock numbers but we know how many and how long we can graze on the summer pasture, and the local government knows that too. However, when there are investigations from the state, we just try to pretend that we don't use the grazing ban zone. We have to implement the rules; no local government dares to reject the policy, because there is no room for that. We say that '*Gon Nas Ckav Yod, Vog Nas Bkod Pa Yod*,'[5] meaning that there are policies from above, and there are solutions from the bottom.[6].

As his quote reveals, policies and fiats from above, are not fixed; rather, there is space to negotiate rules, aspects of which are subverted in accordance with local needs, and sometimes in ways that are accepted by local government. When multiple actors are involved in the governance and control of land and natural resources, the likelihood of negotiations increases. After all, the state–local relationship is also in the process of being made and transformed. As pastoralists improvise orders from the state, the relations with the state and local authorities are also repeatedly tested for the *de facto* gain of the pastoralists.

Rules are deeply rooted with the social, cultural and even emotional dimensions of making them; they are never fixed and are always evolving in a plural context. The case of Saga showed that multiple rules and regulations exist from different power actors toward rangeland control and governance. This multiplicity provides room for flexibility and improvisation on the ground. The notion of legal pluralism highlights the flexibility of adopting rules according to different issues and contexts to meet the gains and claims; and which rules to obey and which regulations to follow is very much the choice of the pastoralist communities. The term 'legality' is fuzzy since the actors and institutions that generate order are always overlapping and, in the process of making it, and the boundaries between legal and non-legal are often blurred. Pluralism, in this sense, underlines how individuals employ multiple rules to justify and legitimize their claims to resource control and access (Meinzen-Dick and Pradhan, 2002). This is akin to 'forum shopping' (von Benda-Beckmann, 2002), when individuals seek out different forums to support their claims, based on which law or interpretation of the law they believe is most likely to support their interests. Thus, while state policies and legal orders generate platforms for powerful actors to engage in resource management, they also expand the room to manoeuvre for less powerful actors that compete for authority under multiple political contexts and institutional landscapes (Sikor and Lund, 2009; Suhardiman, Bright and Palmano, 2021).

In the case of Saga, pastoralists often improvise, reinterpreting rules and regulations that are imposed by local government and the state. For example, the rangeland use regulations were developed in response to state-led scientific ecological research and rangeland conservation concerns, and include the grazing ban and are balanced by the REPSRM.[7] Rather than following the state-designed grazing ban zones based on the monitoring data of vegetation and biomass conditions, pastoralists from Saga designed the grazing ban zone based on the location and distance of the pasture from the government headquarters, as well as the relatively short usage of the summer pasture.

Thus, even though the REPSRM programme determined to restrict the use and access to rangeland through banning grazing and controlling livestock numbers in rural pastoralist regions, the rights to access rangeland and the livestock limit are still determined by the customary village rules and practices, such as *rsta srun*, which Saga pastoralists have coordinated to protect the collective summer pasture and monitor its use and livestock number during the summertime.

Kenya case: practices of assembling open rangelands from above and renegotiating access from below

In the Kenyan context there is a long history of scholarship focused on pastoral institutions (Lesorogol, 2008; Mwangi, 2007). Past institutional analysis has focused on (1) privatization and fragmentation of collectively titled land and the breakdown of resource management due to increasing inequality and state land and agricultural policies (Mwangi, 2007), and (2) critical institutional analysis of wildlife conservation land use on collectively titled land (Bersaglio and Cleaver, 2018; German et al., 2017a). Building on this work, I (second author) focused on how institutions mediate access (Ribot and Peluso, 2003) across 'complex mosaics of property' including private and collectively tenured land (Robinson, 2019), to understand how plural, overlapping rights and authorities, and social processes of inclusion and exclusion all shape access to land (German et al., 2017b; Gongbuzeren, Huntsinger and Li, 2018). Below, I briefly highlight how institutions mediating access to grazing were intertwined with other rules and norms such as market relations, and norms of mutual assistance and reciprocity, and how different actors produce and reproduce asymmetries through their abilities to shape institutional configurations and mediate processes of inclusion and exclusion (Cleaver and de Koning, 2015; Sikor and Lund, 2009).

While previous institutional perspectives have been instructive, my work aimed to add an understanding of socially-differentiated abilities to access and benefit from ecologically variable land, and how this was intertwined with state and non-governmental interventions. As in Tibet, rules of rangeland management have long been 'hybrid' in central and southern Kenya. Case studies from a new institutional economics framing emphasize the function of institutions and show how 'the commons' are sustained through 'customary' institutions that the Maasai use to regulate rangeland access (Mwangi, 2007; Mwangi and Ostrom, 2009). However, the grazing institutions that are sometimes described as 'customary' in this literature have also been heavily shaped by colonial and post-colonial authorities. 'Chiefs' who were not historically Maasai authorities, were leaders selected by the colonial state (Hughes, 2006; Waller, 1976) who played a strong role in designating land use zonation and regimes of grazing access. Elder interlocutors, including 'chiefs', have contested the functionalist logic of grazing regimes intended to regulate pasture, which is sometimes implied in new institutionalist work, and instead pointed to more complex and flexible historical ways of negotiating variability of pastures and key resources (Unks, 2022; Unks et al., 2023). Similarly, anthropological perspectives have pointed to the importance of how shifting systems of patronage and values under colonial and post-colonial regimes became intertwined with rules, norms and values of land (Behnke, 2018; Galaty, 1981). These new rules and norms were followed by the establishment of 'group ranches' (GRs) which established collectively titled land, granted legal authority to elected officials and established new rules and norms of internal range management (Mwangi, 2007; Mwangi and Ostrom, 2009; Rutten 1992) that were given precedence over previous 'customary' institutions (Okoth-Ogendo, 1989). Structural adjustment, constitutional reforms, wildlife conservation reforms and market-based conservation have all had cascading implications for rangeland institutions (Unks, 2022).

I used mixed-methods, including focus-group discussions, semi-structured interviews, informal interviews and surveys, in collaboration with colleagues, focusing on rules and norms of pastoral access to land across collectively titled pastoralist GRs, state and private conservancies, private, and open-access land (see Unks et al., 2019, 2023). Pastoral lands in Kenya are increasingly envisioned from above in a distinct way because of the well-recognized importance of these lands for wildlife mobility outside of national parks. The most salient changes in rules and norms of access during my research were those being negotiated by pastoralists in tension with wildlife conservation interventions, which were building on past institutional changes to reconfigure access. I understood the process of institutional assemblage as diverse actors hybridizing rules and norms that had different origins in long-held Maasai rules and norms, colonial-era indirect rule, post-colonial land governance, state practices of decentralization and structural adjustment, conservation interventions and Maasai social movements to reconfigure land use.

Below I focus on institutional changes due to wildlife conservation interventions, which were dominated by unsteady alliances between transnational NGOs, ecotourism businesses, Kenyan politicians, the state wildlife conservation agency (KWS) and Maasai elites. One key institutional intervention came about in NGOs sustaining existing GR structures (Unks et al., 2023), or supporting their establishment in Laikipia (Figure 11.2; Unks et al., 2019). The GR as a formal system of rules and authority structure in both Laikipia and Kajiado contexts (Figure 11.2; Unks, 2022; Unks et al., 2019) enabled conservation practices to be formally established, through close relationships between NGO and GR representatives. Most conspicuously, the GR structure enabled authorities to establish wildlife conservancies on collectively titled land, which introduced new boundaries and systems of rules that limited access to grazing and water, in contested ways (German et al., 2017a; Unks et al., 2023). In areas where NGOs supported formal titling of GRs, this formally divided contiguous rangelands but also divided land use within GRs to regulate use between areas of settlements, water resources, pastures and conservancies (Unks et al., 2019).

Through legislation in Kenya that permitted the delegation of wildlife management and development practices to NGOs, they were able to develop a system of regulatory and more 'intimate' practices of intervention in institutions. New authorities were established, notably scouts and rangers who, as employees of NGOs, or representatives of local 'community' NGOs that were funded by transnational NGOs, also were in other positions of authority such as GR representatives. As these actors were from Maasai communities, they were able to directly enforce state policies of land use and wildlife management, as well as formal grazing committee rules (Unks, 2022). The increasing presence of NGO representatives and employees from local communities further allowed transnational NGOs to directly influence grazing committee decisions and grazing management bylaws. NGOs advocated openly for leveraging the 'useful' institutions of the Maasai, as seen in both land management plans that NGOs drafted for GRs as well as in scientific publications (Unks, 2022). Additionally, chiefs worked together with NGO employees, grazing committees and GR representatives to recontour land and water access institutions (German et al., 2017a; Unks et al., 2019, 2023).

Under conditions of decreasing access to collective, private and public lands, as well as decreasing labour availability due to state pressures for youth to attend public schools, there was decreasing ability to flexibly respond to forage when it was most beneficial to animal health, and overall longer-distance migrations were limited (Unks et al., 2023). At the same time these new institutions enabled Maasai elites to have greater control over land. Importantly, these institutional changes differentially impacted people who relied on livestock according to wealth, clan and other social connections (Unks et al., 2023). Wealthier herders built assets to enable access to diverse types of land, to engage with markets more beneficially, including through stock brokering, and better able to support pastoralism through diversified incomes (Unks et al., 2019, 2023). This social differentiation was also tied directly to the new benefit streams produced by wildlife conservation and how benefits were distributed through employment and informal systems of patronage (Unks et al., 2019, 2023).

As access to land and indirect financial benefit streams were intertwined with how GR reps influenced institutional configurations and sustained a hybrid tenure regime that enabled disproportionate control over land and indirect benefits, community members responded in a number of ways to influence and contest institutions. Because pastoralists' agency and ability to adapt through local institutions were greatly reduced, they were forum shopping through different institutional levels at clan, GR, county and state levels to seek recourse and contest leaders' claims (see Cleaver and de Koning, 2015). Community members simultaneously made claims to state authorities to contest conservation institutions, to challenge privately held conservation land on settler ranches in Laikipia, and mobilized in diverse ways to challenge GR authority. In Kajiado, Maasai youth and women, in particular, resisted the institutions of collective land management that enabled enhanced accumulation by elites (Unks et al., 2023). In Kajiado, the subdivision of collective land into private land came to be seen by many Maasai as a way of equalizing livelihoods and also wresting power over land from representatives (Unks et al., 2023). In both Laikipia and Kajiado, pastoralists responded by shifting local informal institutions of labour sharing toward waged labour because inequalities in labour sharing allowed elites to take advantage of the labour of less wealthy herders (Unks et al., 2019, 2023). This was closely related to increases in small stock (sheep and goat) production and more sedentary adaptations in response to markets, climate and changing vegetation and adoption of other livelihood practices in both contexts (Unks et al., 2019, 2023). However, the temporary assemblage of GR representatives, Maasai elites, NGOs representatives and politicians – formed around a discourse of lands open to wildlife and livestock, systems of patronage and 'intimate' practices – ultimately enabled disproportionate influence over institutional configurations (Unks et al., 2019, 2023).

Discussion

By examining case studies in pastoral China and Kenya, we observe that institutional systems are always in a state of assemblage through complex relational processes at different levels of social, political and economic organization. Analysing rangeland practices through the lens of assemblage, as introduced, helps in reconciling rangeland policies and practices. In the case of China, access rights to rangeland are still determined by the customary village rules and practices, such as the conventional *rsta srun* practice, where Saga pastoralists have coordinated to protect the collective summer pasture and monitor its use and livestock number during the summertime. The perception of the state controlling, and local actors resisting, is continuously challenged. In other words, control by the state is always limited, and a hybridity of rules, relations and roles are seen. Various arrangements are improvised, reformulated and remade from beneath, not necessarily directly challenging state power (Hall et al., 2014 Ptackova, 2019). Rather it is about finding ways for negotiations and to work in local contexts through practices of assemblage (Heilmann and Perry, 2011; Ho, 2005, 2017; Scoones et al., 2018; Tsering (Huadancairang), 2023; Yeh, O'Brien and Ye, 2013). This is to say that land politics are not restricted to activities within the hands of authorities, whether it is the government or other organizations, but are centred on the fluid practices of negotiations and contestations involving multiple resource user groups.

By taking a critical institutional approach, and focusing on processes of assemblage where diverse actors formed unsteady alliances, the Kenyan case shows how institutions evolved in a context of ongoing political economic changes in land and wildlife policy, changes in Maasai social structure and international wildlife conservation initiatives. This helped to show how institutions were modified in the process of negotiation between diverse actors who were temporarily assembled around a 'governmentality' of institutional reconfiguration (Unks, 2022). However, institutional configurations were co-produced through pastoralists trying to recontour institutions to adapt to constrained mobility, as well as through formal, regulatory and 'intimate' practices used to modify and instrumentalize pastoral institutions for wildlife conservation goals. A key insight that emerged from this approach was an understanding that the challenge of the 'commons' was not so much about how pastoralists regulated variable semi-arid rangeland resources, as it was about the balance of complex competing interests both within and outside of the 'commons' to establish new land uses within collectively titled land. Through a range of practices, top-down influence tended to dominate the process of assemblage and to constrain informal and formal Maasai institutions of land sharing (German et al., 2017a; Unks, 2022). Numerous practices such as physical enforcement of boundaries, practices of disciplining land use and also regulatory measures that work through informal and formal processes, limited pastoralism in its ability to flexibly respond to variability (Unks et al., 2019; Unks, 2022). Most of these institutional changes serve to create new territorial formations where landscapes open to wildlife mobility are emphasized, but pastoralists in turn access land in a highly uneven, socially differentiated and contested way (Unks et al., 2019; Unks, 2022).

Conclusion

A paradigm change is occurring in the study of informal and formal rules and norms of pastoralist natural resource governance. Recent studies of pastoralism have challenged many assumptions about individual and collective behaviour, property regimes, origins and processes of institutional formation, and social power underpinning dominant approaches, leading to a proliferation of new approaches to studying pastoralist institutions.

This chapter highlights how the authors applied alternative approaches in two pastoralist contexts in China and Kenya. The authors consider systems of institutions as always in a state of assemblage through complex relational processes at different levels of social, political and economic organization. This shows how governance in practice involves hybridity between 'customary' practices, political economic processes and national-level land and rangeland management policies and regulations. This also highlights fluidity in practice across governance domains that tend to be treated in isolation analytically.

Acknowledgements

This work was funded by the European Union (ERC, CONDJUST, 101054259). Views and opinions expressed are however those of the author(s) only and do not necessarily reflect those of the European Union or the European Research Council Executive Agency. Neither the European Union nor the granting authority can be held responsible for them. This work also contributes to ICTA-UAB "María de Maeztu" Programme for Units of Excellence of the Spanish Ministry of Science and Innovation (CEX2019-000940-M).

Notes

1 A concept that comes from Foucauldian-inspired feminist theory and refers to the processes whereby subjectivities come into being, and subjectification occurs as power is internalized and re-expressed (Butler 1997; Nightingale 2019). As such, subjectification is a dynamic process that can change over time and space.

2 'A social form shapes the materiality of the matter. It is the shape that a matter becomes if people interact with it in a specific manner' (Euler, 2018: 11).

3 'opportunities for individual growth and self-development are combined with the search for shared solutions, meaningful activities with extended and depend relationships, and the creation of material abundance with the care for others and for nature. Living together like this was and still is practised to various degrees all over the world. In the process, commoning has to be repeatedly scrutinised, updated, and rehearsed in order to remain embedded in everyday life. This can never be taken for granted, and needs a suitable framework which currently we can rarely find' (Acksel et al., 2015: 134).

4 Under the REPSRM mechanism, the core of the programme is to compensate the pastoralists for rangeland where livestock grazing is banned (grazing ban zone), and to reward them for maintaining livestock population within determined rangeland carrying capacity (balanced zone).

5 "གོང་ནས་བཀའ་ཡོད།།འོག་ནས་བགོད་པ་ཡོད།།".

6 Interview, Kokonor, Village Leader, Mar, 2021.

7 '草原生态保护补助奖励机制', *Caoyuan shengtai baohu buzhu jianglijizi*, or Rangeland Ecological Protection Subsidy and Reward Mechanism (REPSRM).

References

Agrawal, A. and Perrin, N. (2009) Climate adaptation, local institutions and rural livelihoods, in Adger, W. N., Lorenzoni, I. and O'Brien, K. L. (eds.) *Adapting to Climate Change*, Cambridge: Cambridge University Press, pp. 350–367. https://doi.org/10.1017/CBO9780511596667.023

Anderson, D. (2002) *Eroding the Commons: The Politics of Ecology in Baringo, Kenya, 1890s–1963*, Melton: James Currey.

Anderson, B. and McFarlane, C. (2011) Assemblage and geography, *Area*, 43(2): 124–127. https://www.jstor.org/stable/41240473

Aneesh, A. (2017) 'Relocating global assemblages': An interview with Saskia Sassen, *Science, Technology and Society*, 22(1): 128–134. https://doi.org/10.1177/0971721817694927

Archambault, C.S. (2016) Re-creating the commons and re-configuring Maasai women's roles on the rangelands in the face of fragmentation, *International Journal of the Commons*, 10(2): 728–746. https://doi.org/10.18352/ijc.685

Bauer, K. and Nyima, Y. (2011) Laws and regulations impacting the enclosure movement on the Tibetan Plateau of China, *Himalaya, the Journal of the Association for Nepal and Himalayan Studies*, 30(1): 10.

Behnke, R. (2018) Open access and the sovereign commons: A political ecology of pastoral land tenure, *Land Use Policy*, 76: 708–718. https://doi.org/10.1016/j.landusepol.2018.02.054

Bennett, A., Acton, L., Epstein, G., Gruby, R.L. and Nenadovic, M. (2018) Embracing conceptual diversity to integrate power and institutional analysis: Introducing a relational typology, *International Journal of the Commons*, 12(2): 330–357. https://doi.org/10.18352/ijc.819

Bersaglio, B. and Cleaver, F. (2018) Green Grab by Bricolage – The institutional workings of community conservancies in Kenya, *Conservation and Society*, 16(4): 467–480. https://doi.org/10.4103/cs.cs_16_144

Butler, J. (1997) *The Psychic Life of Power: Theories in Subjection*, Standford: Stanford University Press.

Cleaver, F. (2001) Insitutional Bricolage, conflict and cooperation in Usangu, Tanzania, *IDS Bulletin*, 32(4): 26–35. https://doi.org/10.1111/j.1759-5436.2001.mp32004004.x

Cleaver, F. (2012) *Development Through Bricolage: Rethinking Institutions for Natural Resource Management*, Abingdon, Oxon: Routledge.

Cleaver, F.D. and de Koning, J. (2015) Furthering critical institutionalism, *International Journal of the Commons*, 9(1): 1–18. https://doi.org/10.18352/ijc.605

Cleaver, F., Franks, T., Maganga, F. and Hall, K. (2013) Institutions, security, and pastoralism: Exploring the limits of hybridity, *African Studies Review*, 56(3): 165–189. https://doi.org/10.1017/asr.2013.84

Côte, M. and Korf, B. (2018) Making concessions: Extractive enclaves, entangled capitalism and regulative pluralism at the gold mining frontier in Burkina Faso, *World Development*, 101: 466–476. https://doi.org/10.1016/j.worlddev.2016.11.002

Dean, M. (2010) *Governmentality: Power and Rule in Modern Society*, 2nd edition, London: SAGE.

Euler, J. (2018) Conceptualizing the commons: Moving beyond the goods-based definition by introducing the social practices of commoning as vital determinant, *Ecological Economics*, 143: 10–16. https://doi.org/10.1016/j.ecolecon.2017.06.020

Fernández-Giménez, M.E. (2002) Spatial and social boundaries and the paradox of pastoral land tenure: A case study from postsocialist Mongolia, *Human Ecology*, 30(1): 49–78. https://doi.org/10.1023/A:1014562913014

Fischer, A.M. (2008) "Population invasion" versus urban exclusion in the Tibetan areas of Western China, *Population and Development Review*, 34(4): 631–662. https://doi.org/10.1111/j.1728-4457.2008.00244.x

Galaty, J.G. (1981) Land and livestock among Kenyan Maasai, *Journal of Asian and African Studies*, 16(1–2): 68–88. https://doi.org/10.1163/156852181X00391

Gelek, L. (2002) The Herders of Eastern Tibet: Notes from an anthropological field study, *Chinese Sociology & Anthropology*, 35(1): 21–81. https://doi.org/10.2753/CSA0009-4625350121

German, L., King, E., Unks, R. and Wachira, N.P. (2017b) This side of subdivision: Individualization and collectivization dynamics in a pastoralist group ranch held under collective title, *Journal of Arid Environments*, 144: 139–155. https://doi.org/10.1016/j.jaridenv.2017.04.009

German, L.A., Unks, R. and King, E. (2017a) Green appropriations through shifting contours of authority and property on a pastoralist commons, *The Journal of Peasant Studies*, 44(3): 631–657. https://doi.org/10.1080/03066150.2016.1176562

Gongbuzeren, Huntsinger, L. and Li, W. (2018) Rebuilding pastoral social-ecological resilience on the Qinghai-Tibetan Plateau in response to changes in policy, economics, and climate, *Ecology and Society*, 23(2): 21. https://doi.org/10.5751/ES-10096-230221

Gruschke, A. (2008) Nomads without pastures? Globalization, regionalization, and livelihood security of Nomads and Former Nomads in Northern Khams, *International Association of Tibetan Studies*, 4: 1–40. https://www.thlib.org?tid=T5570

Gruschke, A. (2012) Tibetan Pastoralists in transition. Political change and state interventions in Nomads, in Kreutzmann, H. (ed.) *Pastoral Practices in High Asia*, Dordrecht: Springer, pp. 273–289. https://doi.org/10.1007/978-94-007-3846-1_15

Hainan Daily (2019) 'Prioritising the Rangeland ecological protection in Hainan TAP'. Hainan TAP Gov. 2016. Implementation plan of a new round of the Rangeland ecological protection subsidy and reward mechanism in Hainan prefecture (2016–2020), *Hainan Daily*. Available at https://www.moa.gov.cn/gk/tzgg_1/tfw/201603/t20160304_5040527.htm. Accessed 19 February 2024.

Hall, P.A. and Taylor, R.C.R. (1996) Political science and the three new institutionalisms, *Political Studies*, 44(5): 936–957. https://doi.org/10.1111/j.1467-9248.1996.tb00343.x

Hall, K., Cleaver, F., Franks, T. and Maganga, F. (2014) Capturing critical institutionalism: A synthesis of key themes and debates, *The European Journal of Development Research*, 26: 71–86. https://doi.org/10.1057/ejdr.2013.48

Helfrich, S. (2012) The logic of the commons and the market: A shorthand comparison of their core Beliefs, in Bollier, D. and Helfrich, S. (eds.) *The Wealth of the Common: A World Beyond Market and State*, Amherst, MA: Levellers Press.

Helfrich, S. and Heinrich-Böll-Stiftung (eds.) (2012) *Commons: für eine neue Politik jenseits von Markt und Staat*, 1. Auflage. ed. Bielefeld: Transcript.

Heilmann, S. and Perry, E.J. (eds.) (2011) *Mao's Invisible Hand: The Political Foundations of Adaptive Governance in China*, Cambridge, MA: Harvard University Press.

Hughes, L. (2006) *Moving the Maasai : A Colonial Misadventure*, St. Antony's Series, Basingstoke: Palgrave Macmillan.

Jeppesen, M.D. and Hassan, R. (2022) Private property and social capital: Dynamics of exclusion and sharing in the subdivided Pastoral Rangelands of Kajiado, Kenya, *Society & Natural Resources*, 35(1): 92–109. https://doi.org/10.1080/08941920.2022.2026542

Johnson, C. (2004) Uncommon ground: The 'poverty of history' in common property discourse, *Development and Change*, 35(3): 407–434. https://doi.org/10.1111/j.1467-7660.2004.00359.x

Latour, B. (2005) *Reassembling the Social: An Introduction to Actor-Network-Theory*, Oxford: Oxford University Press.

Lesorogol, C.K. (2008) *Contesting the Commons: Privatizing Pastoral Lands in Kenya*, Ann Arbor: University of Michigan Press.

Li, T.M. (2007) Practices of assemblage and community forest management, *Economy and Society*, 36(2): 263–293. https://doi.org/10.1080/03085140701254308

Li, T.M. (2014) What is land? Assembling a resource for global investment, *Transactions of the Institute of British Geographers*, 39(4): 589–602. https://doi.org/10.1111/tran.12065

Lind, J., Okenwa, D. and Scoones, I. (eds.) (2020) *Land, Investment and Politics: Reconfiguring Eastern Africa's Pastoral Drylands, African Issues*, Martlesham: Boydell & Brewer. https://doi.org/10.1017/9781787449077

Lund, C. (2006) Twilight institutions: Public authority and local politics in Africa, *Development and Change*, 37(4), 485–705. https://doi.org/10.1111/j.1467-7660.2006.00497.x

Lund, C. (2008) *Local Politics and the Dynamics of Property in Africa*, Cambridge: Cambridge University Press.

Meinzen-Dick, R. and Pradhan, R. (2002) Legal pluralism and dynamic property rights, *CAPRi Working Paper 22*, Washington, DC: International Food Policy Research Institute.

Mies, M. and Bennholdt-Thomsen, V. (2001) Defending, Reclaiming and Reinventing the Commons, *Canadian Journal of Development Studies/Revue canadienne d'études du développement*, 22(4): 997–1023. 10.1080/02255189.2001.9669952

Moritz, M. (2016) Open property regimes, *International Journal of the Commons*, 10(2): 688–708. https://doi.org/10.18352/ijc.719

Mosse, D. (1997) The symbolic making of a common property resource: History, ecology and locality in a tank-irrigated landscape in South India, *Development and Change*, 28(3): 467–504. https://doi.org/10.1111/1467-7660.00051

Mwangi, E. (2007) *Socioeconomic Change and Land Use in Africa*, New York: Palgrave Macmillan. https://doi.org/10.1007/978-1-137-06659-6

Mwangi, E. and Ostrom, E. (2009) A century of institutions and ecology in East Africa's Rangelands: Linking institutional robustness with the ecological resilience of Kenya's Maasailand, in Beckmann, V. and Padmanabhan, M. (eds.) *Institutions and Sustainability: Political Economy of Agriculture and the Environment – Essays in Honour of Konrad Hagedorn*, Dordrecht: Springer, pp. 195–222. https://doi.org/10.1007/978-1-4020-9690-7_10

Nightingale, A.J. (2011) Beyond design principles: Subjectivity, emotion, and the (ir)rational commons, *Society & Natural Resources*, 24(2): 119–132. https://doi.org/10.1080/08941920903278160

Nightingale, A.J. (2019) Commoning for inclusion? Commons, exclusion, property and socio-natural becomings, *International Journal of the Commons*, 13(1): 16–35. https://doi.org/10.18352/ijc.927

Nori, M. and Scoones, I. (2019) Pastoralism, uncertainty and resilience: Global lessons from the margins, *Pastoralism*, 9: 10. https://doi.org/10.1186/s13570-019-0146-8

North, D. (1987) Institutions, transactions costs and economic growth, *Economic Inquiry*, 25(3): 419–428. https://doi.org/10.1111/j.1465-7295.1987.tb00750.x

Nyima, Y. (2021) The implementation of China's Rangeland protection program: The case of Nagchu in the Tibet autonomous region, *Journal of Contemporary China*, 30(132): 977–992. https://doi.org/10.1080/10670564.2021.1893561

Okoth-Ogendo, H.W.O. (1986) *The Perils of land tenure reform: The Case of Kenya*, Nairobi: Faculty of Law, University of Nairobi.

Okoth-Ogendo, H.W.O. (1989) Some issues of theory in the study of tenure relations in African agriculture, *Africa*, 59(1): 6–17. https://doi.org/10.2307/1160760

Ostrom, E. (1986) An agenda for the study of institutions, *Public Choice*, 48: 3–25. https://doi.org/10.1007/BF00239556

Ostrom, E. (1990) *Governing the Commons: The Evolution of Institutions for Collective Action*, Cambridge: Cambridge University Press.

Ostrom, E., Burger, J., Field, C., Norgaard, R. and David, P. (1999) Revisiting the commons: Local lessons, global challenges, *Science*, 284(5412): 278–281. https://doi.org/10.1126/science.284.5412.278

Peluso, N.L. (1996) Fruit trees and family trees in an anthropogenic forest: Ethics of access, property zones, and environmental change in Indonesia, *Comparative Studies in Society and History*, 38(3): 510–548. https://doi.org/10.1017/S0010417500020041.

Peluso, N.L. and Lund, C. (2011) New frontiers of land control: Introduction, *Journal of Peasant Studies*, 38(4): 667–681. https://doi.org/10.1080/03066150.2011.607692

Peluso, N.L. and Ribot, J. (2020) Postscript: A theory of access revisited, *Society & Natural Resources*, 33(2): 300–306. https://doi.org/10.1080/08941920.2019.1709929

Ptackova, J. (2019) Traditionalization as a response to state-induced development in rural Tibetan areas of Qinghai, PRC, *Central Asian Survey*, 38(3): 417–431. https://doi.org/10.1080/02634937.2019.1635990

Rasmussen, M.B. and Lund, C. (2018) Reconfiguring frontier spaces: The territorialization of resource control, *World Development*, 101: 388–399. https://doi.org/10.1016/j.worlddev.2017.01.018

Ribot, J.C. and Peluso, N.L. (2003) A theory of access, *Rural Sociology*, 68(2): 153–181. https://doi.org/10.1111/j.1549-0831.2003.tb00133.x

Robinson, L. (2019) Open property and complex mosaics: Variants in tenure regimes across pastoralist social-ecological systems, *International Journal of the Commons*, 13(1): 805–827. https://doi.org/10.18352/ijc.903

Roche, G. and Hillman, B. (2017) *Why are so many tibetans moving to Chinese cities?* Chinafile. Available at https://www.chinafile.com/reporting-opinion/viewpoint/why-are-so-many-tibetans-moving-chinese-cities

Rutten, M.M.E.M. (1992) *Selling Wealth to Buy Poverty: The Process of the Individualization of Landownership Among the Maasai Pastoralists of Kajiado District, Kenya, 1890–1990*, Nijmegen Studies in Development and Cultural Change, Saarbrücken: Breitenbach.

Sassen, S. (2008) Neither global nor national: Novel assemblages of territory, authority and rights, *Ethics & Global Politics*, 1(1–2): 61–79. https://doi.org/10.3402/egp.v1i1.1814

Schlager, E. and Ostrom, E. (1992) Property-rights regimes and natural resources: A conceptual analysis, *Land Economics*, 68(3): 249–262. https://doi.org/10.2307/3146375

Scoones, I. (2015) *Sustainable Livelihoods and Rural Development*, Rugby: Practical Action Publishing.

Scoones, I., Edelman, M., Borras, S.M., Hall, R., Wolford, W. and White, B. (2018) Emancipatory rural politics: confronting authoritarian populism, *The Journal of Peasant Studies*, 45(1): 1–20. https://doi.org/10.1080/03066150.2017.1339693

Scott, W.R. (2014) *Institutions and Organizations: Ideas, Interests and Identities*, 4th edition, Thousand Oaks, CA: Sage.

Sikor, T. (2008) New publics and property rights, in Sikor, T. (ed.) *Public and Private in Natural Resource Governance: A False Dichotomy?* Abingdon: Earthscan, pp. 215–228.

Sikor, T. and Lund, C. (2009) Access and property: A question of power and authority, *Development and Change*, 40(1): 1–22. https://doi.org/10.1111/j.1467-7660.2009.01503.x

Suhardiman, D., Bright, J. and Palmano, C. (2021) The politics of legal pluralism in the shaping of spatial power in Myanmar's land governance, *The Journal of Peasant Studies*, 48(2): 411–435. https://doi.org/10.1080/03066150.2019.1656200

Topgyal, T. (2011) Insecurity dilemma and the Tibetan uprising in 2008, *Journal of Contemporary China*, 20(69): 183–203. https://doi.org/10.1080/10670564.2011.541627

Tsering (Huadancairang), P. (2023) Tibetan buddhist monastery-based rangeland governance in Amdo Tibet, China, *Land Use Policy*, 131: 106756. https://doi.org/10.1016/j.landusepol.2023.106756

Tsering, P. (2023) Over time and space: Hybrid Rangeland governance in Amdo Tibet China, *Nomadic Peoples*, 27(1): 71–93. https://doi.org/10.3197/np.2023.270105

Turner, M.D. (2011) The new pastoral development paradigm: Engaging the realities of property institutions and livestock mobility in Dryland Africa, *Society & Natural Resources*, 24(5): 469–484. https://doi.org/10.1080/08941920903236291

Turner, M.D. and Schlecht, E. (2019) Livestock mobility in sub-Saharan Africa: A critical review, *Pastoralism*, 9: 13. https://doi.org/10.1186/s13570-019-0150-z

Unks, R.R. (2022) Instrumentalizing pastoralism? Understanding hybrid tenure and governance in Ilkisongo Maasai land of southern Kenya, *Political Geography*, 99: 102712. https://doi.org/10.1016/j.polgeo.2022.102712

Unks, R.R., Goldman, M.J., Mialhe, F., Gunnell, Y. and Hemingway, C. (2023) Diffuse land control, shifting pastoralist institutions, and processes of accumulation in southern Kenya, *The Journal of Peasant Studies*, 50(5): 1757–1790. https://doi.org/10.1080/03066150.2022.2160630

Unks, R.R., King, E.G., German, L.A., Wachira, N.P. and Nelson, D.R. (2019) Unevenness in scale mismatches: Institutional change, pastoralist livelihoods, and herding ecology in Laikipia, Kenya, *Geoforum*, 99: 74–87. https://doi.org/10.1016/j.geoforum.2018.12.010

von Benda-Beckmann, F. (2002) Who's afraid of legal pluralism?, *The Journal of Legal Pluralism and Unofficial Law*, 34(47): 37–82. https://doi.org/10.1080/07329113.2002.10756563

von Benda-Beckmann, F., von Benda-Beckmann, K. and Eckert, J. (2009) Rules of Law and Laws of Ruling: Law and Governance between Past and Future, in von Benda-Beckmann, F., von Benda-Beckmann, K and Eckert, J. (eds) Rules of Law and Laws of Ruling: On the Governance of Law, Farnham: Ashgate, pp. 1–30.

Velicu, I. and García-López, G. (2018). Thinking the Commons through Ostrom and Butler: Boundedness and Vulnerability, *Theory, Culture & Society*, 35(6): 55–73. https://doi.org/10.1177/0263276418757315

Waller, R. (1976) The Maasai and the British 1895–1905 the origins of an alliance, *The Journal of African History*, 17(4): 529–553. https://doi.org/10.1017/S002185370001505X

Waller, R. (2012) Pastoral production in Colonial Kenya: Lessons from the past?, *African Studies Review*, 55(2): 1–27. https://doi.org/10.1353/arw.2012.0039

Waylen, G. (2014) Informal institutions, institutional change, and gender equality, *Political Research Quarterly*, 67(1): 212–223. https://doi.org/10.1177/1065912913510360

Whaley, L. (2018) The critical institutional analysis and development (CIAD) framework, *International Journal of the Commons*, 12(2): 137–161. https://doi.org/10.18352/ijc.848

Yeh, E.T., O'Brien, K.J. and Ye, J. (2013) Rural politics in contemporary China, *Journal of Peasant Studies*, 40(6): 915–928. https://doi.org/10.1080/03066150.2013.866097

Yeh, E.T., Samberg, L.H., Gaerrang, Volkmar, E. and Harris, R.B. (2017) Pastoralist decision-making on the Tibetan Plateau, *Human Ecology*, 45: 333–343. https://doi.org/10.1007/s10745-017-9891-8

12 Conclusion

Fiona Nunan

Introduction

The preceding chapters drew on a diverse range of theories and approaches to research institutions in the context of natural resource governance. In so doing, they introduced, reviewed and reflected on a number of frameworks, theoretical perspectives and data collection and analysis methods. The purpose of this chapter is to bring together key findings and reflections from across the chapters to identify key insights gained on institutions and what lessons were shared regarding researching institutions in natural resource governance. Table 12.1 sets out key points and findings from each chapter, together with key lessons, insights and tips for researching institutions.

Before reviewing these, it is worth reflecting on how the chapters understood institutions and what the focus of research tended to be. All chapters recognised both formal and informal institutions to at least some degree, acknowledging that both are generally present and can be researched using the methods and approaches introduced and reviewed. In several cases, formal policies and legislation were researched, but collective rules and local norms were also recognised and investigated. Chapter 3 demonstrated how Institutional Grammar can be used to identify whether an institutional statement is a rule, norm or strategy, and, in that way, demonstrates what an institution sets out to do, that is, its scope or remit. Chapter 4 provided a different perspective on examining or measuring the performance of institutions by recognising the different dimensions of institutions, that is, their content, input, activities, outputs and outcomes. Focusing on one component, or perspective, of an institution can facilitate the study of an institution in practice, given that the performance and effectiveness of institutions can be challenging to identify. These chapters provided a way of categorising institutions, with relevance to both formal and informal institutions.

Many of the chapters reported on using the framework, approach or methods to investigate which institutions were present, how important participants felt institutions were and to measure or assess the effectiveness of institutions. Both quantitative and qualitative methods were used to assess effectiveness, or perceptions of effectiveness, with some approaches enabling both types of

DOI: 10.4324/9781032624679-12

Table 12.1 Methods and lessons

Chapter	*Framework/approach and focus of research*	*Research design and methods*	*Key lessons, insights and tips*
2	Institutional Analysis and Development (IAD) framework Action situations: communal assembly meetings and collective lands Research investigated: Institutional change Institutional interaction Influence of institutional change and interaction on behaviour and outcomes Attributes of the communal governance system	Comparative case study design Two stage approach: **Stage 1:** structured interviews with community leaders of communities participating in Payment for Ecosystem Services scheme and leaders of non-participating communities **Stage II:** quasi-experimental design – participating and non-participating communities Data gathered at two points in time, 2013 and 2018: Focus group discussions carried out before household questionnaire to inform questions Household survey Rapid land use assessment and visual analysis of satellite images to triangulate data from surveys Analysis: Logit regression models Difference in Difference Descriptive statistics Qualitative analysis of FGDs and open-ended questions	1 IAD provides an organising structure to research institutions in complex socio-ecological systems 2 Identifying institutions can be challenging as not all written down Institutional Grammar can help by categorising rules Also used protocols from the International Forestry Resources and Institutions to inform phrasing of questions http://ifri.forgov.org/ 3 Many institutions and explanatory variables could be investigated – inevitably hard decisions need to be made – trade-offs between depth and breadth Undertaking research in stages and using both qualitative and quantitative methods can help with this

3	Institutional Grammar (IG) Embedded within the IAD framework Enables analysis of institutional statements and so also the examination of the institutional structure of action situations Multiple institutional statements for institutional arrangements	Involves coding of phrases and terms of institutional statements, to identify patterns in information and terms used Descriptive analysis – May involve qualitative and quantitative analysis – Used to identify types of institutional designs and categories of institutional arrangements – Can be used to distinguish between rules, norms and strategies – Can also be used to test hypotheses regarding institutional designs Statistical analysis – Provides a descriptive design of IG data in numerical terms – Statistical modelling can facilitate examination of factors contributing to observed to institutional statements Network analysis and modelling – Institutions can be treated as networks of prescribed interactions – Can use standard network analysis measures, such as centrality and density, to explain institutional statements Computational analysis – Agent-based modelling has particularly been used in relation to IG – IG can inform the design of models or to explain behaviour patterns derived from the models	1 IG has proven to be an effective analytical approach in institutional analysis 2 Can be used to analyse many forms of institutions, enabling differentiation between types of institutions 3 Challenges have arisen related to coding, but standardised approaches exist 4 However, recognition of the context of institutional statements remains

(*Continued*)

Table 12.1 (Continued)

Chapter	*Framework/approach and focus of research*	*Research design and methods*	*Key lessons, insights and tips*
4	Logics modelling for quantitative measuring of institutions	Five types of measures of institutions: 1 Institutional content: may involve applying typologies and coding of institutions 2 Institutional input: often involves assessing the human and final resources allocated 3 Institutional activities: for example, meetings, patrols, monitoring, conflict resolution 4 Institutional outputs: refers to the immediate results or products associated with institutional activities 5 Institutional outcomes: the functional contributions of the institutions Measuring institutions may draw on many sources, often for the same study, such as government policies and legislation, public and organisational records, surveys, interviews and participant observation	1 Measuring institutions is desirable as a way of understanding the effects of institutions 2 There are many reasons why measuring institutions is not straightforward, but key amongst those reasons is that the rules observed in practice are not necessarily what is written down There may be no or incomplete implementation, and differences in interpretation and application 3 Each type of measure provides an angle on the performance of institutions but also has limitations
5	Fuzzy cognitive mapping A participatory modelling method involving creation of structured causal maps with stakeholders Can use maps to identify institutions and influence on behaviour and outcomes, comparing perspectives and scenarios	FCM may take a positivist or interpretivist approach, generating objective or subjective knowledge to predict or explain institutional influence and change Generate mental models via causal maps in either small groups or with individuals; case study example used focus groups of less than six people, all from the same stakeholder group Groups enables deliberation but individual mapping may be better for sensitive issues Coded qualitative data from interviews/discussion Established a common terminology across the multiple maps, integrated these into a mental modelling software, aggregated maps within groups, compared across groups and finally produced a 'crowd' map based on causal maps from all group discussions Conducted metric analysis and scenario analysis	1 Level of participation of stakeholders can vary based on many factors, including research purpose, design and time and resources available 2 Power relations and representation in groups will affect dialogue and maps Need to consider inclusion, including in relation to gender, language, knowledge, etc 3 Epistemological assumptions shape the knowledge shared and how it is interpreted and used 4 FCM inevitably involves simplification and so care needs to be taken in interpretation, recognising that some knowledge may be prioritised above others

6	The complex adaptive governance systems framework (CAGS-F) Provides a model of social complexity across levels and scales of complex socio-ecological systems	Used a case study approach with field observations, interviews and document review to identify relevant institutions, organisations and human practice; iterative processes of data collection and analysis could be useful Analysis includes clustering individuals into groups based on internal homogeneity and external heterogeneity of social practices, with race, class or gender, or dominant ideologies, considered after initial clustering Data from media, advertising, photographs or written texts can be drawn on to inform understanding of norms, values, ideologies and relevant contexts	1 CAGS-F was informed by scholarship on institutions, organisations, intersectionality and panarchy heuristic within a critical realist orientation 2 Recognition of the reality and value of diversity is central to the framework, as is the pursuit of justice 3 Intersectional and stratification theories facilitate understanding, and increase inclusion, of diverse realities and differential capacities 4 Critical realism brings out invisible forces and relations of power and privilege
7	Network analysis Social network analysis: whole system network Policy network analysis: actor-centric network in relation to a specific topic/ issue Three levels of analysis: 1 network configurations: how actors are situated in governance network and patterns of social influence 2 actor positions 3 community sub-groups	Methods could include: interviews, surveys, document analysis, longitudinal approaches, participatory methods and mixed-methods Interviews and participatory methods may involve mapping a network with interviewees; interviews and surveys may involve using name-generator questions, followed by name interpreter questions, collecting attribute information on the names or information on the ties Network analysis often involves the use of specialised software and statistical software	1 Networks provide a visualisation of relationships between actors, which can include institutions, and enables analysis of factors influencing those relationships, or ties 2 Networks can be used as a boundary object to facilitate collaborative analysis 3 Can demonstrate the complexity of networks and different perceptions of actors, their attributes and ties between respondents 4 Network analysis can be used to identify which actors and ties could drive change or where there are gaps in connections between actors/institutions

(Continued)

Table 12.1 (Continued)

Chapter	*Framework/approach and focus of research*	*Research design and methods*	*Key lessons, insights and tips*
8	Ethnography Unstructured, flexible approach, gathering data on everyday functionings	Very often involves observation – participant or non-participant, where the researcher immerses themselves in the research situation, almost mirroring 'natural conditions' Immersion particularly useful for identifying socially embedded institutions that respondents might not have thought of as relevant Range of other methods used: semi-structured interviews, focus group discussions and document analysis Example reported on also involved household surveys, carried out twice in a relaxed ethnographic way, feeding into the other data collection methods Range of data collection methods used to map institutions and then investigate them more deeply	1 Very suitable for researching evolving institutions, such as power relations and kinship 2 A gatekeeper may be necessary to introduce a researcher, so implications of this need to be considered and, if appropriate, responded to, eg by building trust over time 3 It is challenging to research some institutions as they operate in intangible spaces and activities relating to institutions may be sporadic 4 Ethical concerns may be raised by some approaches, for example undertaking covert observation so that those being observed do not alter how they would behave without being observed 5 Data recording can present challenges, particularly during observation Taking rapid notes during and immediately after observation is one solution, with more detailed notes made later

9	Socio-technical Infrastructure as an object of study Sociotechnical tinkering as an entry point	Ethnographic immersion – including semi-structured interviews (several rounds of interviews), participant observation, diary keeping Participatory mapping and transect walks Symmetrical analysis	1 The materiality of infrastructure matters in institutional functioning and change 2 Positionality matters in researching institutions and implementing technography 3 Researching institutions is always experimental and can never fully be planned in advance There is need for flexibility and innovation
10	Practices of assemblage and actor network theory Assemblage as a 'dynamic convergence of human and non-human actors, their actions and their environments, in complex webs of relations (Chapter 10, p. 173)	Recommends steps to go through: 1 Identify the 'rules that matter', rather than being guided only by written rules 2 Map the functioning and evolution of institutions 3 Trace the causes of institutions 4 Identify and illustrate the behavioural significance of institutions Used historical and ethnographic research methods, starting with a review of secondary sources before interviewing all actors and observers found to be relevant, from Indigenous Peoples groups to forestry bureaucrats and eco-lodge entrepreneurs Observation enabled through participation in biological monitoring through a student internship scheme	1 Practice matters more than written rules, so start with practice 2 Bricolage between and of institutions is inevitable 3 Institutional analysis must be aware of the influence of power asymmetries and politics 4 Resource governance is not static, but fluid and multifaceted 5 Qualitative methods are more appropriate than quantitative for researching the evolution of institutions

(Continued)

Table 12.1 (Continued)

Chapter	*Framework/approach and focus of research*	*Research design and methods*	*Key lessons, insights and tips*
11	Assemblage 'the practice of bringing together an array of agents… and objectives' (Chapter 11, p. 189) Notes close connection between assemblage and critical institutionalism	Ethnographic case studies, involving semi-structured in-depth interviews, key-informant interviews, participant observation, informal interviews/back stage talks/casual chat and photovoice	1 Taking an assemblage approach helps to identify resource governance practices and, through this, institutions 2 An assemblage approach recognises that institutions are continually assembling, drawing on heterogeneous elements 3 Finds that institutions themselves are 'inherently political, social, relational entities that are undergoing constant change' (Chapter 11, p. 189) 4 'Rules are not fixed or settled; they are challenged and reinterpreted to suit local realities and needs' (Chapter 11, p. 189) 5 Institutions are inevitably entwined 6 Case studies demonstrated how institutions introduced through top-down interventions interact with emerging bottom-up practices

methods, such as Fuzzy Cognitive Mapping and network analysis. The review of five measures of the effectiveness of institutions in Chapter 4 included the identification of limitations associated with different ways of measuring effectiveness, recognising challenges in delivering consistency between studies. In Chapter 2, measuring the effectiveness of institutions was facilitated by data collection at two points in time, as well as by the use of several methods, which enabled the effectiveness of institutions to be measured, or assessed, in different ways.

Whilst, then, the chapters cover a wide range of theoretical perspectives, informed by contrasting philosophical underpinnings on causality and reality, they share an endeavour to understand the purpose, practice and outcomes of institutions. To examine the key learning points in more detail, the chapter first identifies what is learnt about institutions from the different approaches to researching institutions. This is followed by a section on what can be learned about researching institutions in natural resource governance and, finally, by identification of key tips for researchers, drawing together key points from across this chapter.

What do we learn about institutions?

Most of the key findings in relation to what we learn about institutions in relation to natural resource governance are already known. They are worth restating, however, as recognition and understanding of these key findings enable implications for researching institutions to be identified. These key characteristics of institutions are that: context matters for institutions; institutions very often look different in practice to what has been written down (where institutions are written down); there are very often many, and diverse, institutions within a given setting; and, finally, that given the number and diversity of institutions present, institutions may interact, influence each other and change as a result.

In considering why context matters for institutions and for researching institutions, it is worth going back to the definitions of institutions given in Chapter 1. The definitions considered institutions as 'humanly devised constraints' (North, 1990: 3) and as

> enduring regularities of human action in situations structured by rules, norms, and shared strategies, as well as by the physical world. The rules, norms, and shared strategies are constituted and reconstituted by human interaction in frequently occurring or repetitive situations.
>
> (Crawford and Ostrom, 1995: 582)

Drawing on these definitions, and on typologies of institutions, whether formal and informal, bureaucratic and socially-embedded or another typology, we know that institutions may be deliberately designed and negotiated, but may also emerge through human interaction. Institutions are therefore found in government policy and legislation, but also in local rules and norms, power relations and kinship. Given the diversity of the form and origin of institutions, it is apparent that there are many components of 'context' that may influence what an institution looks like, how it is interpreted and how it may change over time.

In Chapter 2, Hayes and Murtinho adapt the Institutional Analysis and Development framework to the situation of community-based Payment for Ecosystem Services schemes and identify the relevance of biophysical conditions, community attributes and attributes of governance systems in researching institutional change and the influence of institutions on behaviour. Inspired by the same framework, in Chapter 3, Siddiki and Schlager acknowledge the relevance of context in interpreting institutional statements using Institutional Grammar and, in Chapter 5, context is clearly relevant in May's complex adaptive governance systems framework (CAGS-F) through recognition of the diversity of identities and interests that may be present in, or are affected by, governance systems. An ethnographic approach enables a researcher to be immersed within 'context', as shown by Datta in Chapter 8, and an assemblage approach brings attention to the relevance of non-human as well as human context, as shown in Chapters 10 and 11 by Orihuela, and Tsering and Unks respectively, demonstrating that non-humans have agency as well as being part of the context of institutions. Tsering and Unks in Chapter 11 observed how context led to institutions being reconfigured, including policies and practices of land acquisition, commercialisation, wildlife conservation and carbon sequestration.

Many of the chapters noted that institutions very often look different in practice to what has been written down, meaning that whilst it may be appropriate to research institutions through studying policies and legislation, very often it is not sufficient to stop there. The research reported on in the chapters demonstrated that institutions may be ignored, not known about, interpreted in different ways, shaped deliberately by actors and may interact with other institutions. Researching institutions in practice implies that researchers should acknowledge that institutions may change over time and be interpreted, used and shaped differently between locations and instances. Orihuela in Chapter 10 advises that practice matters more than written rules and so should be the focus of research. Interviews, focus group discussions and observation can be used to generate data on 'institutions in practice'. Mapping approaches were also found to enable the identification of institutions in practice, whether through Fuzzy Cognitive Mapping or network analysis.

Attention to institutions in practice is also necessary as some institutions, particularly informal, or socially-embedded, institutions, can be hard to identify and research, as noted in Chapter 1. The elusiveness of some institutions to a researcher new to a setting was demonstrated in Chapter 8, where Datta emphasised the benefits of immersion in a community to identify 'institutions in practice' that potential respondents might not think of mentioning in surveys, interviews or discussions.

A further rationale for giving attention to institutions in practice is because institutions and their influence on behaviour and outcomes may change over time. In Chapter 11, Tsering and Unks observe that institutions are constantly assembling, are influenced and manipulated by actors, and alter through interaction between institutions. Studying change in institutions and their influence was demonstrated in Chapter 2 by research that involved collecting data at two points in time. This approach enabled Hayes and Murtinho to study how institutions were evident in practice over time, but also whether and how people's behaviour had changed over time, and with what implications for the ecological system.

The number and diversity of institutions present within a given setting means that institutions may interact, influence each other, and change as a result. Interaction between institutions may also be deliberate, as institutions are consciously pieced together to suit or meet an actor's interests or needs, as discussed in relation to 'institutional bricolage' in Chapter 1. The existence of multiple rules in any one context was argued by Tsering and Unks in Chapter 11 as providing 'room for flexibility and improvisation', enabling people to pursue their interests whilst negotiating the institutional landscape, moving between, and piecing together, institutions to enable certain decisions and outcomes.

What do we learn about researching institutions?

Within the approaches and frameworks applied in the chapters, a range of data collection methods were used, including document analysis, surveys, interviews, focus group discussion, participatory mapping and participant observation. In Chapter 2, Hayes and Murtinho demonstrated how rapid land use assessment and visual analysis of satellite images can complement and triangulate data from surveys related to how people are using and managing land and so generate data on the performance of institutions. Several chapters drew on ethnography, itself a broad term, but characterised by research that is relatively open-ended, unstructured and small-scale, taking everyday contexts as the research environment (Hammersley and Atkinson, 2019). These characteristics make ethnography, and the key method associated with it of participant observation, particularly relevant for researching institutions in practice, enabling observation of behaviour rather than generating data on knowledge, perceptions or reported behaviour. Chitata et al. in Chapter 9 report on how observation enabled data collected through other methods to be triangulated by observing practice and comparing it to reported behaviour. Chitata et al. also note how participant observation gave the first author a deeper appreciation of some of the experiences shared by interviewees, for example the drudgery and hard labour involved in cleaning the water night storage tank. The technography approach adopted for the research reported on in Chapter 9 encouraged greater attention to the water irrigation infrastructure and through that, greater engagement with operating and maintaining that infrastructure and hence appreciation of the experience and views of farmers using the infrastructure. Observation can be a challenging method to carry out well, presenting challenges in terms of what to observe, how much to observe and how to interpret observations. Generating data from multiple sources and using multiple methods can assist in responding to these challenges.

A wide range of methods can, therefore, be used to generate data on institutions, informed by the questions being asked, the cases being examined and the interests and skills of the researcher. In particular, the diversity of methods referred to across the chapters demonstrates how the choice of data collection methods to be used may depend on the level of detail, or richness of data, sought versus breadth of coverage of institutions, taking a more quantitative approach with a larger sample.

Five key lessons emerge about researching institutions from the chapters: that often multiple, or mixed, methods are appropriate for researching institutions; that researching institutions can take time; there may need to be trade-offs in generating data; consideration should be given to inclusion in the design and use of methods; and, that qualitative methods may be more appropriate for researching institutional interaction and change than quantitative methods.

The use of multiple, and mixed, methods to research institutions responds to the limitations of relying on one method, given that institutions can be difficult to identify and understand, and people may have different perceptions and experiences of the same institution. Using different methods enables data to be generated on different aspects or experiences of institutions. The term 'mixed methods' refers to the use of both quantitative and qualitative methods, whereas multiple methods could be used, relying on only qualitative or quantitative. Mixed methods research has been defined by its characteristics as research that:

> collects and analyzes both qualitative and quantitative data rigorously in response to research questions and hypotheses, integrates (or mixes or combines) the two forms of data and their results, organizes these procedures into specific research designs that provide the logic and procedures for conducting the study, and frames these procedures within theory and philosophy.
>
> (Creswell and Plano Clark, 2018: 31)

Mixed-methods research enables the strengths of both qualitative and quantitative research to be benefited from, with weaknesses offset. There are many typologies of mixed methods designs, with Creswell and Plano Clark (2018) identifying three core designs from a review of typologies, being: convergent, where the results of qualitative and quantitative methods are brought together or compared; explanatory sequential, which begins with quantitative data collection and analysis feeding into more in-depth explanation through qualitative data generation and analysis; and, exploratory sequential, which begins with qualitative data collection, feeding into the design of quantitative data collection and analysis to test the generalisability of findings from the qualitative phase. Which design is chosen will depend on the research questions being asked and the knowledge the researchers have of the empirical situation.

The use of a mixed, two-phase approach to researching institutions reported in Chapter 2, demonstrated how data generated through one method can inform the design of subsequent data collection involving different methods. Similarly in Chapter 8, Datta used a survey in a context of ethnographic research as a way of generating a broad-based understanding of institutions in relation to forest management, but also to introduce herself to the study communities and begin building trust. The use of a survey enabled the identification of aspects of institutions that could be investigated in more detail through observation and interviews, taking an explanatory approach.

Given that institutions can be challenging to identify and researching institutions often requires qualitative approaches to generate rich accounts of the nature and experience of institutions, it can take time to research institutions. A further reason for taking time concerns the need to build trust in the researcher and the research by people involved in the setting/case. Building trust is essential for meaningful engagement in research, for participants to open up and share insights that enable deeper analysis of institutions. Trust is also essential when people are being observed for research purposes, so that they behave as they would if they were not being observed. To assist with the amount of time needed, and building trust, some researchers reported going via a gatekeeper to access relevant people and communities. In Chapter 8, Datta reflects on the pros and cons of being introduced by a Forest Guard to her case study communities, as the Guard provided legitimacy, by indicating that the government accepted her as a researcher, but also recognised that being associated with the government may affect how people behaved with her, at least initially. Accessing participants via a gatekeeper can assist in reducing the time taken to undertake research, but needs to be undertaken with care, given the potential distrust in institutions and the existence of power asymmetries.

Given that studying institutions can take time, limited resources, including time, may lead to trade-off decisions needing to be made between breadth and depth, as discussed in Chapter 2. Hayes and Murtinho observe that decisions may need to be made between focusing on one, or a smaller number of communities/sites, to generate in-depth, fine-grained analysis of institutions, and generating data from a larger number of sites, to generate evidence on broader patterns associated with the nature and performance of institutions.

A further key lesson learnt from the chapters on researching institutions is that in many cases, a range of people should be interviewed and observed, recognising that people have different experiences and knowledge of institutions. Such different experiences may themselves be shaped by institutions, such as gender norms and power relations. Given power differentials between potential participants in research, it can be difficult to ensure that respondents have the space and confidence to share their views and experiences. This is particularly the case in group discussions, as found in the case reported in Chapter 6, where Fuzzy Cognitive Mapping involved the generation of causal maps. Power differences may lead to some group members speaking more than others or their contributions carrying greater weight, particularly where consensus is needed on which institutions to record and how much weight to assign to influence behaviour and outcomes. Strategies to enable fair and meaningful inclusion are needed for informed, rigorous research. The CAGS-F in Chapter 5 explicitly enables analysis of diversity in relation to the governance of social–ecological systems, recognising that power differences associated with class, caste, race, income disparities, etc. are relevant to who is involved in, and affected by, decision-making.

Finally, several authors made the case that qualitative methods are more suited to the study of evolving and interacting institutions than quantitative methods. Orihuela (Chapter 10) and Tsering and Unks (Chapter 11) argued that assemblage provides an appropriate lens to study the dynamics and complexity of institutions, given that institutions are in a 'state of flux' (Chapter 11). Institutions were found to be assembled and reassembled over time, requiring attention to how institutions are interpreted, reinterpreted and pieced together to suit objectives and context.

Observations on researching institutions in natural resource governance

From the above review of what has been learnt about institutions, and researching them, in relation to natural resource governance from the chapters, the following observations can be discerned:

1 Research into institutions involved in, or influencing, natural resource governance, whichever methods or frameworks are used, or questions asked, is likely to involve, and generate data on, both formal and informal institutions. It is evident from the chapters that it is hard to avoid consideration of both forms of institutions, given their prevalence and interaction. However, research, and certain frameworks and methods, may give greater attention to one form of institution above the other. Awareness of how different types or forms of institutions interact, complement or conflict with each other is therefore needed in researching institutions.
2 When researching formal institutions in relation to natural resource governance, what is found in practice may differ to what has been written down. And what is found in practice may change over time and may be different between locations at the same time. In addition, people within a setting may have different perceptions and experiences of institutions. They may, therefore, rank or score the influence or effectiveness of institutions differently. Interactions between institutions may partly explain why and how the practice, experience and perceptions of institutions differ over time and space.
3 Given the diversity of experience and effectiveness of institutions, the context of institutions matters and should, therefore, be taken into consideration in research. Taking context into consideration may be through the research design and sampling, or through generating data on attributes

of the context to feed into the analytical framework, model or qualitative analysis.

4 There are many methods, approaches and frameworks that may be used to research institutions in natural resource governance, with selection informed by the context and purpose of the research, as well as the resources and skills available. Very often, multiple or mixed methods may be used, reflecting the need to examine institutions from different perspectives, triangulate data and generate data on different aspects of institutions.

5 It can take considerable time to research institutions, reflecting challenges in identifying institutions and understanding their origin, purpose and effectiveness. Given the time researching institutions in natural resource governance can take, trade-off decisions may need to be taken, between in-depth research into one or a small number of cases, and broader research with a larger sample and/or range of data being generated. To address the potential challenge that taking considerable time may present, researchers may consider seeking a gatekeeper to ease access to research participants, using a multi-stage approach to inform decision-making on methods and facilitate effective use of resources, and use multiple or mixed methods to generate data from different perspectives and, potentially, sources.

6 Given that the nature, sharing and influence of power are integral to natural resource governance, power dynamics are also found in institutions in a number of ways. Power asymmetries may be reflected in the purpose, nature and outcomes of institutions, and institutions may, therefore, perpetuate power asymmetries. Researching institutions in natural resource governance should, therefore, pay attention to the existence and implications of power asymmetries. Such power asymmetries may affect how institutions are discussed and practiced, and therefore people's experience and perceptions of them. Careful consideration is, therefore, needed of the actors invited to participate in research and how research is conducted, particularly in group settings, where some voices may be louder, or more evidently heard, than others. There are then two dimensions to recognising and responding to power asymmetries in researching institutions in natural resource governance. The first is to discern how institutions may reflect, reinforce and perpetuate power asymmetries in natural resource governance and, therefore, include analysis of the nature and manifestations of power using an appropriate conceptualisation, typology or framework. Bennett et al. (2018) provide a typology of interactions between power and institutions, asking how institutions shape power and how power shapes institutions that can guide such research. Partelow and Manlosa's (2023) investigation into power and commons governance can also inform research into power and institutions in natural resource governance. And the second dimension is to consider how inclusive research processes are, including recruitment of participants, the timing of data collection, where data collection takes place and how participants may influence the contributions of others in group situations.

7 Whilst institutions do limit uncertainty in interactions and, in the context of natural resource governance, guide decision-making, it is also clear that they can and do change over time, in response to many factors. Researching institutions, therefore, should be aware of their dynamic nature, aiming to capture how an institution may have changed or evolved over time through questions asked of participants, undertaking longitudinal research or modelling change in institutions by incorporating a range of evidence. The dynamics of institutions may be associated with contextual change, interaction between institutions and by the re-interpretation and reformation of institutions through interactions between people, and between people and non-human actors. Certainly, the dynamics of institutions can make researching institutions in natural resource governance challenging, but being aware of the potential for change, in the past and in the future, is critical for effective research.

The chapters in this book do not provide an exhaustive account of the methods, frameworks and approaches that can guide research into institutions associated with natural resource governance, neither are they exhaustive in relation to the experience of using these. However, the diversity of methods, frameworks and approaches, and reflections on their use have generated consistent and informative insights into how institutions have been, and can be, researched. It is hoped that the observations listed above will guide and inspire researchers in designing and undertaking research into institutions in the context of natural resource governance, whether using the methods, frameworks and approaches included in the book or others. It is also hoped that this guidance and inspiration will lead to a deepened understanding and appreciation of institutions, and, in so doing, contribute to more effective, fair and sustainable natural resource governance.

References

Bennett, A., Acton, L., Epstein, G., Gruby, R.L. and Nenadovic, M. (2018) Embracing conceptual diversity to integrate power and institutional analysis: Introducing a relational typology, *International Journal of the Commons*, 12(2): 330–357. https://doi.org/10.18352/ijc.819

Crawford, S.E.S. and Ostrom, E. (1995) A grammar of institutions, *The American Political Science Review*, 89(3): 582–600. https://www.jstor.org/stable/2082975

Creswell, J.W. and Plano Clark, V.L. (2018) *Designing and Conducting Mixed Methods Research*, 3rd edition, Thousand Oaks, CA: Sage.

Hammersley, M. and Atkinson, P. (2019) *Ethnography: Principles in Practice*, Abingdon: Routledge.

North, D.C. (1990) *Institutions, Institutional Change and Economic Performance*, Cambridge: Cambridge University Press.

Partelow, S. and Manlosa, A.O. (2023) Commoning the governance: A review of literature and the integration of power, *Sustainability Science*, 18: 265–283. https://doi.org/10.1007/s11625-022-01191-2

Index

Note: Bold page numbers refer to tables.

FSC
www.fsc.org
MIX
Papier | Fördert gute Waldnutzung
FSC® C083411

Kolibri 360 GmbH
Ferdinand-Jühlke-Straße 7
99095 Erfurt, Deutschland
produktsicherheit@kolibri360.de